THE NATURE (
Spirit an

Publisher's Note

This book was first published in German in 1950 and in English
translation in 1966 (with a second edition in 1983). Given the many
decades that have passed since it was written, the text would clearly
benefit from a thorough checking and re-editing. Without doubt,
some of the research on which Hauschka has based his ideas is out of
date, and the scientific context has shifted. Nevertheless, the book
contains an enormous amount of groundbreaking work that remains
highly relevant to the present day.

Given the fact that we are not in a position to fund a major reworking
and redesign of the text—and given the many requests we have
received for the book over the years—we have taken the decision to
publish a facsimile of the 1983 English edition, and trust that readers
will continue to find it of value.

THE NATURE OF SUBSTANCE

Spirit and Matter

Rudolf Hauschka

SOPHIA BOOKS
Rudolf Steiner Press

Sophia Books
Hillside House, The Square
Forest Row, East Sussex
RH18 5ES

www.rudolfsteinerpress.com

Published by Sophia Books 2002; reprinted 2008
An imprint of Rudolf Steiner Press

First published in English by Stuart & Watkins 1966
Translated from German by Marjorie Spock & Mary T. Richards
Originally published in German by Vittorio Klostermann,
Frankfurt am Main 1950

A catalogue record for this book is available from the British Library

ISBN 978 185584 122 2

Cover by Andrew Morgan Design, based on design for previous
edition by Derek Ritterband Studios
Printed and bound in Great Britain by 4edge Limited

Contents

CONTENTS

Foreword to the First English Edition (1966)

This book is an attempt to show that an appropriate study of the nature of matter itself will overcome the materialistic view of nature now held almost everywhere.

It may appear a contradiction in terms to speak of a non-materialistic science of chemistry; does not a branch of knowledge that deals with the nature of substances necessarily base itself on the laws of matter?

It is far from the purpose of this work to deny their validity. Indeed, material phenomena and the laws that govern them will form the point of departure for our treatment. We shall, however, pay the closest attention to the limits within which any fact obtains – something which present-day science all too easily neglects.

The author has been at work for decades on experiments which have yielded a new perspective on the nature of matter and hence suggest a new orientation of the sciences. But eyes that educate themselves to qualitative seeing need no experiments to view long-familiar phenomena in a fresh light.

Overcoming a materialistic view of nature means learning to see phenomena freshly, practising a way of looking that is an active 'thinking into' things – not merely a recording of measure, weight and number, followed by explanations that imprison facts in a rigid world of hypotheses and theory.

A former materialistic emphasis on naturalism in German art led perceptive observers to give a warning against this trend as ruinous to painting and sculpture. They pointed out that, by naturalistic standards, colour photography could be called the peak of possible artistic achievement rather than simply the marvel of technology it is.

An artist who merely copies nature, no matter how perfectly, is a technician, not an artist. Genuine artists *live* in the objects of their study and create them freshly and revealingly again. This type of creating calls for something that goes beyond trained hand or eye: it requires vivid activity of soul and spirit. The warmer and more intensive this becomes, the more creative is the artist. So he penetrates ever closer to the heart of truth, out of which the reality of outer nature also sprang. Is there any reason why science could not benefit equally from this artistic approach to facts? In our view it should, and by so doing satisfy the justified demand of the human spirit for creativity in all its functioning.

To record what has been observed is in itself neither a way to truth nor even genuine experience. And for our purposes theories are worse than useless. The world of ideas which comes to light in man must be brought to bear on his perceptions if he is to achieve real knowledge. A creative element, an active conscious-ness, gives itself to the world in a way that makes of every per-ception a living whole, bearing the stamp of essential experience.

Goethe said: 'All experience is necessarily produced, brought forth, created.' Speaking of his journeys to Italy, he remarked that though he had seen nothing new, his *way* of seeing was itself new.

The public has long given evidence of a wish and need for an approach to chemistry imbued with Goethean liveliness. The author feels impelled to try to meet this need. Especially the English-speaking countries seemed to challenge him to help in laying the basis for a premonitory experience of the spirit behind nature. As a student of Rudolf Steiner he embarked on a Goethean line of research that led to a picture of the universe by no means at odds with the orthodox one, but rather tending to round out and complete it.

The book first appeared in German. The public responded with a degree of interest that has so far led to three German editions, as well as to two related books on nutrition and healing.

The following thoughts were uppermost in my mind as I pursued the goal described above. My presentation was to be kept general and easy to understand, without sacrificing scientific character. Hence technical terms were avoided as far as possible, along with the standardized style common to scientific works. On the other hand, I was concerned to capture the reader's immediate interest for my subject-matter with vivid descriptions that would bring home to him how alive, how relevant, it really is. In this connection I want to thank the translators, Marjorie

Spock and Mary T. Richards, for preserving the life and colour of the style I developed.

My thanks are also due to the publisher, Vincent Stuart, for the interest shown in making possible and preparing this English edition of my book.

RUDOLF HAUSCHKA, D.Sc.

Boll, June 1965

Foreword to the Second English Edition (1983)

Since this book was first published, over thirty years ago in Germany, and some twenty years ago in English translation, interest has grown significantly in scientific views, like those of the late Dr Rudolf Hauschka, that directly challenge the assumptions of materialistic science. This volume is being reprinted, with minor corrections, with the purpose of stimulating this interest further.

It must be borne in mind, however, that since the time when the text was written, research in biological chemistry and physics, notably vitamins (see Chapter XV), has progressed enormously, and many basic questions have been resolved.

The main reason for delaying the reprinting of this volume derives from uncertainties about, and lack of independent corroboration of, the various experiments described. It has not proved possible to repeat them based simply on the all too brief and generalized descriptions given in the text, or so far to trace the original experimental documents.

While this may detract from the scientific merit of the book, it must be appreciated that, on the other hand, it has not been demonstrated conclusively that the experiments cannot be repeated in their original form. The subtle and delicate effects within the realm of the four ethers cannot be so readily demonstrated in laboratory experiments as purely inorganic or electromagnetic forces can. The Science Group of the Anthroposophical Society in Great Britain is continuing to seek to clarify this vital issue.

Hence the Science Group considers that the reprinting of this volume meanwhile is justified subject to this written qualification, and that many of Dr Hauschka's imaginative insights and seminal ideas contained herein are in themselves a most valuable contribution to our scientific understanding of chemistry and the living world. The Group welcomes any written experimental evidence, either for or against the subject matter of this book, in particular published reports or articles. Correspondence should be addressed in the first place to the Secretary of the Group, Rudolf Steiner House, 35 Park Road, London NW1 6XT.

June 1983   ROBERT KERSEY GREEN

CHAPTER ONE

Theories of the Natural Scientific Age on the Primality of Matter

K nowledge of matter has grown incredibly in the past centuries. It is natural to ask what called forth this sudden, avalanche-like progress in the exact sciences. If one considers such heroes of recent centuries of science as Lavoisier, Berzelius, Avogadro, Liebig, Whöler and others, one concludes that problems of matter were never studied with such power of observation and logic. One gets the impression that this surprising state of affairs is due to mankind's having reached a new stage of consciousness. The dawning of this change in outlook began to be noticeable in the fifteenth, sixteenth and seventeenth centuries, in the persons of Galileo, Newton and Kepler. Experimental research started to limit itself ever more narrowly to weighing, measuring and counting. A quantitative emphasis laid hold of science. But while facts learned from experimenting were distilled into conclusions that were supposed to permit no excursions outside the realm of the visible, these went more and more in the direction of theories and hypotheses that could not be physically proved. The result was a quantitative and mechanistic world-conception based on hypothesis and logic.

Haeckel's researches and Darwin's theories of descent fitted exactly into this materialistic picture of the world.

It has become a matter of course to think of matter in terms of atomic structure. The prevailing view is that matter is built of atoms, and we assume that atoms, or even more primary particles, are eternal. The law of the conservation of matter has been regarded as one of the most fundamental laws of nature. What facts was it based on?

It was Avogadro who discovered that hydrogen and oxygen always combine in the same ratio.

Two parts of hydrogen plus one part of oxygen equal two parts of water-vapour. Amounts of hydrogen or oxygen in excess of this ratio remain unchanged in the reaction. Further experiments showed that these simple combinations can be doubled or otherwise multiplied, giving rise to the law of simple and multiple proportions. Manganese, for example, combines with oxygen in the following ratios:

$$1:1 \quad Mn + O = MnO \text{ (manganese oxide)}$$
$$2:3 \quad 2Mn + 3O = Mn_2O_3 \text{ (manganese sesquioxide)}$$
$$1:2 \quad Mn + 2O = MnO_2 \text{ (manganese dioxide)}$$
$$1:3 \quad Mn + 3O = MnO_3 \text{ (manganese trioxide)}$$
$$2:7 \quad 2Mn + 7O = Mn_2O_7 \text{ (manganese heptoxide)}$$

The logical mind now reacts as follows: according to Avogadro, two litres of hydrogen combine with one litre of oxygen. Similarly, two cc. of hydrogen combine with one cc. of oxygen, and two cmm. of hydrogen with one cmm. of oxygen. This process may be followed to the point of indivisibility, in which case two indivisible quanta of hydrogen combine with one indivisible quantum of oxygen – i.e., two atoms of hydrogen combine with one atom of oxygen.

Since it has been found that two litres of hydrogen weigh twice 0·09 grams, and a litre of oxygen 1·43 grams (approximately sixteen times as much), we can say that two grammes of hydrogen plus sixteen grammes of oxygen add up to eighteen grammes of water. These weight ratios came to be termed atomic or molecular weights. This lent the atom a dubious reality and endowed it with a permanence of sorts. Avogadro's glorious rhythmic-musical, dynamic law becomes rigidified for no good reason into a spatial-material conception – that of a world composed of atoms. What began as a numerical relationship became a fixed and static picture of substantial particles.

It is not hard to see how the popularizing of these theories and conclusions could give rise to a world conception based on an assumed indestructibility of matter. But all serious scientific research, too, was built on the same assumption. Kant and Laplace brought out their theory of a primal nebula, composed of primal matter. This nebula was assumed to contain all the atoms of which our present earth and universe are built.

Past decades did not regard this conception as problematical, but they did have trouble explaining how life could possibly have originated in such a cosmos. Many theories were advanced, with the consensus that life must have resulted from a chance but complex constellation of atoms.

According to Haeckel and Darwin, the life thus produced went on and developed into ever new and different forms, to the point of creating a nervous system and a brain that became an organ able to produce what we call psychic and spiritual functions. These ideas are not restricted to scientists; in the course of time they become common property, and every layman was picturing a universe supposedly built of pre-existing matter.

The discovery of radium and the study of phenomena associated with it came close to shattering this world conception, for it was found that radium did not obey the law of the conservation of matter. It disintegrates into electricity, warmth, light and various substances such as lead, helium and other elements. Belief in the indestructibility of the atom was rudely shaken by these findings. But radiology, as developed by Lord Rutherford and Bohr, adopted the view that matter is composed of still smaller particles than atoms. The atom has a nucleus, possibly charged with positive electricity (protons and neutrons, the protons being charged with positive electricity), which is surrounded and circled by units called electrons.

Radium gives off three kinds of rays, known as alpha, beta and gamma rays. Alpha rays are composed of particles of matter originating in atomic nuclei. Beta rays consist of electrical particles, electrons. Gamma rays are electro-magnetic radiations, similar to those of light but of much shorter wave-length. Under certain conditions alpha particles can produce warmth. The disintegration of matter into higher than material components such as warmth, light and electricity was explained by postulating nuclei and electrons. Theories of the atomic structure of matter and the conservation of mass were thus able to celebrate their triumphant vindication on a new level.

Further weight was lent to the atomic-electronic view by discoveries of which Planck's quantum theory was one – though actually it brings nothing more than the rhythmical quality of matter and material processes to expression.

These developments were greeted by many devoted university research men with the greatest enthusiasm, for they hoped to find in them clues to the solution of the riddles of man and universe. But they often found, after a lifetime of the most devoted effort, that the path they were following led to a very one-sided truth. Despite its grandeur, this cosmos of atoms simply could not provide the sure footing the soul of man needs in order to take its proper place in the world order. To have their ardent crusade for truth end in the resigned conclusion that human beings were after all nothing more than a chance product of swirling atoms and electrons seemed to rob human nature of its dignity. This dignity requires that at least equal attention should be paid to that missing aspect of the truth which modern science has exiled to the spheres of philosophy or religion.

As things stood, single realms of knowledge had inevitably to grow ever further apart, whereas they should have joined forces to create a harmonious picture of the world in which each shed light upon the others.

CHAPTER TWO

Ancient Ideas about the Nature of Substance

People seem generally inclined to regard early concepts of the nature of matter as primitive and childish. Indeed, we are particularly proud that present-day culture and knowledge have advanced so far beyond the past in this respect. But anyone who gives serious study to ancient civilizations by way of existing writings, buildings and sculpture is amazed at the wealth of wisdom and skill they evidence.

Take, for example, the Egyptian pyramids and temples. They reveal, apart from their obvious artistic merits, such wonders of mathematical and technological capacity as to make it impossible to call their creators primitive or childlike. They contain pillars, columns and statuary hewn out of solid blocks of granite weighing as much as twenty-five tons. Modern engineering, for all its fine technical equipment, would not find it at all easy to handle or transport such gigantic blocks. And the astounding fact is that there are no granite quarries anywhere near these pyramids and temples. The closest one is at Aswan, approximately six hundred miles further up the Nile. To realize that the granite blocks were nevertheless transported all that distance is indeed thrilling. It seems clear that the Egyptians must have possessed capacities which we have lost.

Is there not ample reason to review our belief that man has developed in all respects from a primitive animal-like condition to the heights of our present scientific age? Is it really so naive to credit the possibility that ancient peoples possessed powers which we have lost and must now try to recover with the right use of our intelligence?

There is indeed every reason to think that the men of pre-historic times lived in a very different state of consciousness and were carried along to a great extent by cosmic forces, which endowed them with strength and wisdom far beyond our own. We can recognize that evolution not only brought man his new capacity for thinking, but with it the decline of other powers he once possessed. It is certain that earth-dwellers of the period

5

following what is called the Flood by some, and by others the end of the Ice Age, had a totally different consciousness from ours. Man of that time had no intellectual capacity whatever. He possessed instead the power to perceive a supersensible world of which modern man is unaware.

We are speaking here of a time prior to the writing of the Vedas, when the original Aryans brought the first cultural impulse from the mountains of Central Asia to what is now called India. The supersensible world and its inhabitants were as real to people of that period as is the physical world to present-day humanity. But then the realms of pure spirit were everything, the earth and its kingdoms negligible, unreal – 'Maya' or illusion, the least significant aspect of creation. The continent of Asia was the 'lowest heaven'. Self-awareness of the kind we have today was alien to these early Indians. They felt themselves one with the divine, instruments through which the breath of heaven pulsated. This was the time recorded in the Vedas, though the account, like other such mythological traditions as are found in the Edda and the Kalevala, was written down only at a much later period. They indicate that the ancient Indians had no concept of personal freedom, in the sens.: that they were incapable of making decisions based on their own judgment. They felt themselves supported and permeated by the divine. Not only were the thoughts they had more in the nature of what we would call dreams, but their deeds were the expression of the divine will, not of their own.

Later periods of culture were characterized by an awakening interest in the earth and in man's natural surroundings. Man descended step by step into matter. The early Persians, for example, developed the rudiments of agriculture and began to cultivate the soil. Zarathustra, who was recognized throughout this epoch of history as its teacher, taught his people how to breed the food plants that still form our chief source of nourishment. The Zend-Avesta may be called the first agricultural textbook.

As interest in earthly matters increased, contact with the divine world waned. It would, however, still have seemed very strong by present standards, continuing as it did for some time to be the decisive guide in all man's earthly activity.

In the following cultural period only a select few, such as Egypt's priest-kings, were still able to receive divine illumination and to transmute it into earthly action. The capacity to do so was carefully nurtured in regulating time of birth, special educational measures, and subjection to a rigorous training in the Mystery schools.

This development continued through the Grecian period. Spiritual perception grew ever fainter, even in the case of trained initiates.

Today all contact with the spiritual world is lost. All that remains of it is a dim memory recorded in various religious writings, myths, sagas, fairy tales and dreams. Faculties such as 'second sight' and the like, still occasionally encountered among primitive peoples and even here and there in Europe, are the last vanishing remnants of a once sublime view of the spiritual facts underlying all material phenomena. These faculties are atavistic and out of keeping with the times.

A review of these great developments indicates that divine forces which once approached man from without, from spiritual heights, and which he obeyed unquestioningly, underwent a transformation into powers later found within the soul of man himself, powers to be developed and guided by his own initiative and judgment. The ability to think, self-awareness, and individual freedom were among their fruits, acquired at the cost of sacrificing heavenly wisdom, the fruit of earlier clairvoyance.

Traces of this development are to be found on every hand. At one time the social order, communal life and relationships were formed according to the will of the gods, clairvoyantly explored. The peoples of ancient India and Persia sought revelations of the divine will and obeyed them implicitly. Priests of the Chaldeans and the early Egyptians read in the stars what the gods ordered, and directed every phase of the common life accordingly. Even in historical times Greek clairvoyant observation noted and recorded how the Erinyes or Furies persecuted those who had acted in defiance of the will of the gods. Modern man no longer perceives Erinyes, but he does sense in his heart the bad conscience that reproaches him for letting down his standards. Anyone who reads Homer's account of how Pallas Athene approaches Achilles and whispers in his ear what he should do, or other recitals of heroes being guided by the gods, can scarcely avoid the impression that the Trojan war was really a war of spiritual beings who simply used men like puppets to fight their battles on the earth.

The differentiated macrocosmic forces which worked in and through man, and which he experienced as separate divinities, were as though 'turned outside in' in the course of evolution, to reappear within the human soul. They became man's own activity, instead of continuing to affect him from without. This change did not, of course, occur at the same time in all parts of the world. Greece was the place where it was most conspicuous.

Plato, who experienced his ideas as spiritual visions, may be regarded as the last initiate of early times. The harmony of the spheres was not an abstraction in his mind or just a convenient means of drawing contrasts between cosmic law-abidingness and earthly licence; it was real spiritual music. In his view, the world of matter evolved out of a higher, spiritual realm.

This is the quintessence of many old myths which picture the creation of the world. We will not go further into this here, but merely point out the significant fact that the concept of a gradual densification of matter even found its way into Greek philosophy. This descent of more ethereal elements (fire, light and air) was thought of as a suffusion of matter with spiritual elements, and it was experienced in sublime imaginative pictures. But Plato must have begun to feel that the old delineations were no longer adequate. His pupil, Aristotle, was, in fact, the first to try to cast this ancient visionary wisdom, of which only fragments still remained, in the mould of logical ideas. This is the clue to Aristotle's teaching, and most especially to his doctrine of the elements.

In the light of spiritual history it can be seen that when Aristotle speaks of air he means more than the mixture of oxygen, nitrogen and other gases which we moderns have in mind. His concept was a much broader one that embraced the active forces in which air and all the gaseous elements originate. When he spoke of water he was not referring to the 'H_2O' of modern chemistry, but to the whole third phase of material creation, the fluid element and everything contained within it, including chemical activity. Fire, air, water, earth: these were the milestones in the great process of material evolution.

This life-permeated outlook had to be sacrificed in the course of the following two millennia to an abstract view which enabled thinking to become an independent faculty, free from the influence of the earlier visionary cosmic knowledge. All that was left were the four purely physical 'aggregate states' of matter, with heat soon to be dispossessed of its reality. The world of qualities and action once inseparable from concepts of the elements had to sink out of sight, so that the mind could be schooled exclusively in sense-perception.

Aristotle's doctrines remained the basis of knowledge until well on in the Middle Ages. But they grew ever less lively, ever more material. This had to be so if thought was to come to grips with lifeless matter and develop to the high point it has reached today.

Though this was the general trend of evolution, there ran alongside it a thin thread of another kind of knowledge about nature's secrets. This took into account forces which the new trend disregarded. We are speaking of genuine alchemy.

The prevailing concept of alchemy is most inadequate and misrepresents the facts. Alchemists are usually considered charlatans, ignorant people subject to all sorts of superstitions, men with the one object of turning base metals into gold. There were certainly some charlatans and dilettantes among them, especially towards the end of the period referred to, who earned this bad reputation. But true alchemists such as Basil Valentine, Agrippa of Nettesheim, Raymond of Sabunda, and certainly also Paracelsus, revealed a shining, cosmic knowledge, recognizable as a last spark of the ancient enlightenment.

Much intensive, penetrating study is required for any real understanding of what genuine alchemy was. And students must take into account the fact that writers of alchemistic works usually veiled their meaning; readers must often look for the key to decipher what their words are intended to convey. Today, that wisdom of not so very long ago, which embraced cosmos, earth and man, is almost wholly forgotten.

Just as the ancient clairvoyant vision was crystallized in the

Aristotelian teachings, so for the alchemists the spiritual forces at work in nature were crystallized in the physical and chemical phenomena brought to light by their experiments. These experiments were questions put by them to the goddess Natura, and her answers were regarded with deep reverence. To be sure, the alchemists had no specialized knowledge of the chemical structure of matter such as we have today, but in its place was an awareness of the matrix of earthly and universal forces from which matter sprang and in which it was embedded.

There is a widespread supposition that modern chemistry had its roots in alchemy and has evolved out of the ancient 'primitive' ideas entertained by alchemists. This is incorrect. Alchemy was rather the last fruit of a glorious past. This accounts partly for the decadence into which it finally fell. Modern chemistry is quite the opposite: a new departure in the search for knowledge.

CHAPTER THREE

New Ideas on the Primality of Spirit

In order to discover those aspects of truth which have been lost to modern science, it is by no means necessary to attempt to recover a long-past wisdom in its old form. That would be impossible without resort to an old-style faith. The spirit of the times quite properly demands that knowledge take the place of faith. So the question arises: are there any modern means whereby the truth contained in myths could be made available to science? Would it at least be possible to achieve an inner conviction that reality consists of more than the physical senses can perceive, and to do so in a way that could satisfy a scientific conscience?

The old clairvoyant faculties had been transformed into intellectual capacity. Macrocosmic forces that once directed man from outside had given way to forces awakening in man himself. But does the development of the critical intellect mean that the spiritual evolution of man has come to an end? Or is it rather the modest start of a new era? Does it not seem that the seeds of new human faculties are beginning to germinate?

Although Goethe's work was done more than a century ago, it must be said that his insights into life's deeper laws were wider than those mostly to be found in present-day science. Materialists would deny this, but it often takes centuries to understand genius. Goethe's neglected theory of colour points to the emergence of the new faculties we mean, as do some of his other basic scientific ideas.

As the great opponent of Newton, Goethe attacked Newton's purely intellectual explanation of light with an acerbity that was rare in him. Newton believed that the light-waves he discovered were light itself, or produced light, and this, together with his assumption that light was composed of the various colours, roused Goethe to the most vehement protest. In his view, light-waves were simply the physical manifestation of eternal indivisible light, which produced the spectrum in co-operation with an equally real force of darkness that is something quite different from the mere absence of light. Just as the human body described by anatomy is only the physical aspect of man's eternal being, so, in Goethe's view, was light a far loftier element than the sphere in which its waves were manifest. He spoke of light's moral qualities, of its 'deeds and sufferings' as the origin of colour.

In Goethe, a comprehensive vision was renewed. His observations were invariably directed toward the whole of Nature. He refused to separate the individual phenomenon from its related background. A dynamic rather than a static thinker he thought in terms of polarities and metamorphoses. To his study of Nature's physiognomy, as one might call it, he brought all his resources of sympathetic insight, and so Nature revealed to him more of her secrets than weighing, counting and measuring can ever discover. If we see a smile on a friend's face, it would seem strange to take his blood-pressure and make a urinalysis to ascertain its cause. The cause is immediately apparent when we look into his face with sympathetic sensitivity. That is Goethe's way with nature – his 'power of judging by looking'.

Goethe's time harboured more such men than is commonly realized, thinkers who opposed with all the strength of their idealism the materialism threatening every branch of science. And they brought this to expression not only in art and philosophy, but also in science through the new seeds sown by Goethe and his followers.

The so-called 'Enlightenment' was widely regarded as an attack on the deepest elements in human nature; and to give these a voice in all fields of science was felt to be a task for German culture. So Novalis, in his essay, 'Christendom or Europe', could write the thought-provoking sentence: 'The product of the modern way of thinking was called philosophy and included everything opposed to the old. Fantasy and feeling were proclaimed heresies, along with love of art and morality, past and future. Man was reduced to a mere creature of nature, ruled by animal necessity. The eternal creative music of the universe became the monotonous rattling of a giant mill driven by the stream of chance and floating on it – a mill sufficient unto itself, without maker or miller, a real perpetuum mobile, with only itself to grind.'

These words describe a school of thought limited to but one

aspect of the world, the mechanical. The Goetheanists tried to counter it with a living concept whereby they hoped to learn more of nature than merely its physical-material precipitate. Their goal was to understand life in man and universe, right through to the stage of sense-perceptible phenomena.

Goethe's followers were at pains to carry his method of investigation further. One of their number, the philosopher Preuss, taught that matter and spirit are a unity. According to him, matter is spirit on another level. His essay, 'Spirit and Matter', points to the experiments of Baron von Herzeele of Hanover, whose work, *The Origin of Inorganic Substances*, seems to offer proof that matter is continuously created in the living plant.

In this and later works Herzeele published some five hundred analyses indicating an increase in the potash, magnesium, phosphorus, calcium and sulphur content of seeds sprouted in distilled water only. During the experiments the seeds were placed in porcelain bowls and covered with glass bells equipped with air filters to keep out dust. The law of the conservation of matter would have led one to expect to find exactly the same mineral content in plants grown in distilled water as was found in the seeds from which they developed. But Herzeele's analyses indicate a definite increase not only of mineral ash but of every one of its components.

Herzeele then carried out a further series of experiments, using salt solutions of a pre-determined make-up in place of the distilled water medium. He found, for example, that seeds grown in a solution containing phosphorus in a certain density lessened the solution's phosphorus content. But the seedlings themselves showed no increase in phosphorus. Instead, their sulphur content rose materially. This, says Herzeele, seems to indicate that plants can make sulphur out of phosphorus. He likewise found that plants growing in a nutrient solution of salts of calcium increased their phosphorus content. Moreover, their calcium content increased when they were grown in a magnesium-salt solution. To stimulate magnesium production, he found he had to add carbonic acid. In this particular experiment the seedlings were grown on wire netting woven of platinum in chambers provided with certain quantities of carbonic acid. The vessels were kept moistened with distilled water.

Space permits giving only the following examples as illustrations of his extensive findings.

MAGNESIUM CONTENT WITH NO INTRODUCTION OF CARBONIC ACID

	Before the experiment	After the experiment
I	0·007%	0·022%
II	0·008%	0·036%

WITH CARBONIC ACID

	Before	After
I	0·012%	0·062%
II	0·014%	0·110%

Herzeele believed that his experiments demonstrated the existence of a chain of genetic relationships from carbon dioxide via magnesium, calcium and phosphorus to sulphur, thus:

$$CO_2 \to Mg \to Ca \to P \to S$$

In other experiments he established the formation of potassium from nitrogen:

$$N \to K$$

Thus it appears not only that plants can transform substances, but that the creation of basic elements of matter is commonplace in the organic kingdoms. Herzeele goes so far as to say that dead matter is never of primary origin. 'What lives may die, but nothing is created dead.' And 'the soil does not produce plants; plants produce soil'.

Preuss speaks of these experiments as follows: 'Herzeele's experiments offer tangible proof that the supposed immutability of chemical elements is a fiction that must be speedily discarded if natural science is to progress.'

The tragic thing was that Herzeele's writings, published between 1876 and 1883, were given the silent treatment and lost sight of. Only one copy seems to have survived. Considering the fact that this was a time of many great inventions and discoveries, when, for example, Liebig and Wöhler were developing theories that led to atomistic explanations of biological phenomena, it is perhaps understandable that Herzeele was not given a hearing. It can be stated, after a decade's research by the author, that by and large Herzeele's findings are scientifically sound and by no means as fantastic as they first appear. Many of his experiments were checked and his claims substantiated. An increase in mineral content was definitely established in many cases. But something else turned up of which no mention is made in Herzeele's writings: decreases of mineral content in some other cases. His findings must therefore be extended to the statement that plants not only generate matter out of a non-material sphere, but under certain circumstances again etherealize it.

Herzeele's research did not answer the question whether new matter is actually formed, or whether carbonic acid and nitrogen are simply metabolized into mineral components of the plant. My own research shows plainly that we are dealing here with a fresh creation of substances.

Experiments with seedlings were conducted by the author, substituting air-tight glass vessels for Herzeele's open saucers. Later on, sealed ampoules were used to prevent the entrance or escape of carbon dioxide, nitrogen and other elements. The jars and ampoules were placed on an analytical scale for observation.

If plants generate matter, it was to be expected that the vessel with the seedlings in it would grow heavier, for weight is an attribute of matter. If, on the other hand, matter in plants can actually be etherealized, the vessel of seedlings could be expected to grow lighter.

A detailed account of the experimental set-up and my findings will be published later. However, a short interim report should be presented here.

It is probably unnecessary to emphasize that every possible precaution was taken to obtain scrupulously accurate measurements. Beginning in 1933, a Kaiser and Sievers scale, model PbPll, equipped with projection readings and blower and capable of determining weights up to 0·01 mg., was at my disposal. Accurate readings were obtained with the help of a compensation device: by using a weighing vessel of equal volume

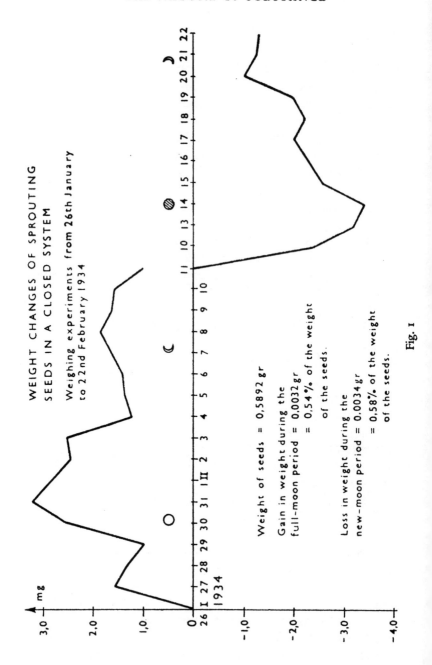

WEIGHT CHANGES OF SPROUTING
SEEDS IN A CLOSED SYSTEM

Weighing experiments from 26th January
to 22nd February 1934

Weight of seeds = 0,5892 gr

Gain in weight during the
full-moon period = 0.0032 gr
= 0.54% of the weight
of the seeds.

Loss in weight during the
new-moon period = 0.0034 gr
= 0.58% of the weight
of the seeds.

1934

Fig. I

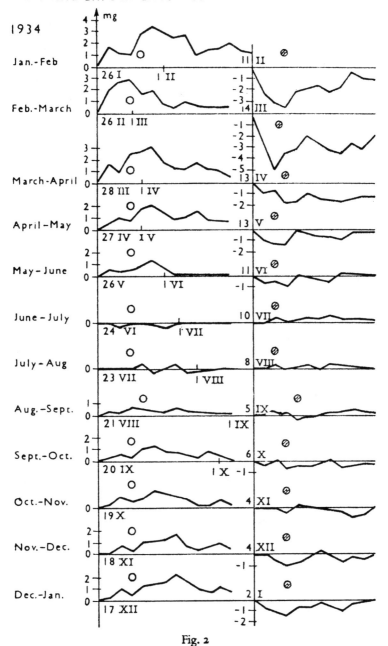

Fig. 2

Weight changes of sprouting seeds in a closed system.

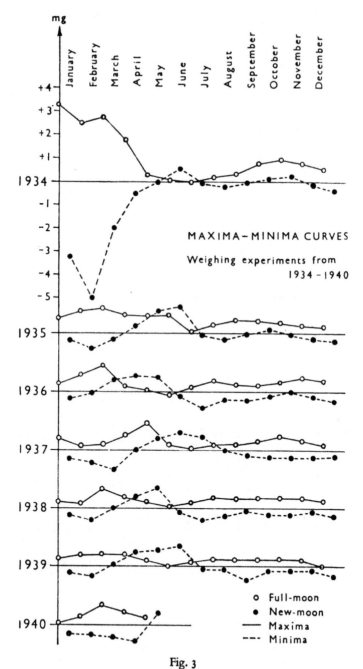

Fig. 3

Weight changes of sprouting seeds in a closed system.

18

as a counter-weight, no allowance had to be made for changes in barometric pressure, temperature or humidity. The weighing vessels had closely fitting lids, firmly cemented on with Ramsay fat. Later on, ampoules of 20 c.c. capacity were substituted for the jars and sealed up after filling.

The margin of error was determined by compensation weighing, comparing an empty tared jar of identical volume with the counter-weight ordinarily employed. It never exceeded ± ·001 mg. Experiments were limited to a duration of fourteen days, as this sufficed for undisturbed growth under reasonably normal conditions.

Measurements now clearly established both increases and decreases in weight, to a degree far exceeding the determined margin of error.

Increases and decreases in weight – or, to put it another way, the emergence and disappearance of matter – occurred in rhythmic sequences which proved the time factor to be the determining one. Figure 1 shows graphs of two successive experiments, with time (i.e. individual days, from January 26, 1934 to February 22, 1934) shown on the horizontal, weight changes in milligrams on the vertical arm. The starting weight is plotted as 0.

The first experiment established a gain of 3·2 mg. between January 26 and February 11, a period during which the moon was waxing. The second experiment, which began on February 11 and continued while the moon was waning, showed a loss of 3·4 mg. Identical or similar results were obtained in subsequent experiments during the years from 1934 to 1940. During this time, experiments consistently gave evidence of weight gains or losses in every waxing and waning moon period respectively, with the findings confirmed by setting up at least two, and sometimes as many as four, simultaneous tests.

Figure 2 gives a survey of experiments performed in 1934. It shows clearly that the moon rhythm is subservient to a stronger annual, or sun, rhythm. Strangely enough, at the year's halfway point in summer, the powerful dynamics of the curves come to a standstill.

Figure 3 shows annual curves for the seven years from 1934 to 1940, in the form of maximum and minimum curves which result from plotting points of maximum and minimum weight changes registered during the full-moon and new-moon periods covered by the experiments. The striking decline in the curves after the great surges of 1934 cannot be discussed here. But it is obvious that a still greater rhythm overrides the annual or sun rhythm just as the latter dominates the rhythm of the moon.

The study of plants involves processes emancipated from the sphere in which mechanical and chemical laws obtain – processes subject rather to cosmic laws and influences.

A further series of experiments performed by the author indicates that even mineral matter reflects planetary influences in physical and chemical reactions. The broad field pioneered by the capillary-dynamic method of research shows definite connections between earthly substances and planetary happenings. The experimental procedure used consists in letting uncoated filter paper absorb solutions by means of capillary attraction. Forces active in a given substance are revealed by this method to a degree unmatched by physical and chemical analysis.

Admittedly, the results of such researches are not readily expressed in terms of measure, weight and number, and a feeling for quality is essential to a grasp of the phenomena observed. Our time is unfortunately still inclined to consider real only what can be weighed, measured and counted. This means still giving Newton precedence over Goethe.

However, Herzeele's and my own subsequent researches substantiate the statement that the law of the conservation of matter applies only to the mineral kingdom and within certain limits, and may not simply be extended to the organic kingdom. We have no right, therefore, to project present conditions of matter into past and future infinities of time or space. Rather is there every reason to assume that matter is the precipitate of life.

Is it not reasonable to suppose that life existed long before matter and was the product of a pre-existent spiritual cosmos? Does it not seem high time to counter the dogma of matter's pre-existence with the idea of the pre-existence of spirit?

CHAPTER FOUR

The Plants

In the rainbow, Goethe recognized the activity of two polar opposites, light and darkness, which bring their yellow and blue to harmonious resolution in the colour green. There is a reconciling element in this many-hued bond between heaven's light and earthly darkness.

The same polar principle is found in plants. Their roots are trapped in the earth's dark depths, while their blossoms reach toward the sunlight. In the plant's green foliage the two processes are harmonized. Indeed, it is here in the leaf that we find the primal form and essence of the vegetable kingdom.

We know that starch is a product of the plant's process of assimilation. This process takes place in the plant's middle zone, the leaf, when sunlight acts upon it in the presence of water and carbon dioxide. Plant physiologists express this in the following formula:

$$6CO_2 + 5H_2O = C_6H_{10}O_5 \text{ (starch)} + 6O_2$$

This rigid atomistic formula and its corollaries fall far short of giving an adequate picture of the miraculous creation of this virgin substance, and they certainly do not explain it. What account do they take of light and of its interplay with darkness?

Nature often gives sensitive observers certain insights. An ocean traveller who has experienced a monsoon (actually a storm in sunlit weather), and has seen the ship enveloped in foam and spray and the sunlight striking through it, making the sky a glorious rainbow dome, can derive some impression from such a phenomenon of the way living substance comes into being. The elements that compose this natural drama are light, air and water. Air-permeated water or water-permeated air, with the sun's rays striking through them, make the rainbow. Light, air and water are the elements that make colours play around fountains and waterfalls. But one can stare and stare at open secrets without fathoming their meaning. For again it is the same elements – light, air and water – which bring about assimilation and go to form starch in plants.

Of course, the airy element in assimilation is 'heavy air', or carbon dioxide, rather than the ordinary atmospheric air. Carbon dioxide is the heaviest gas known to nature; it can be poured, exactly like a liquid, from one container into another. The so-called 'Dog Grotto' in Naples is simply a subterranean lake of carbon dioxide which suffocates dogs that enter it. A man can wade safely through this invisible shallow sea because his head is on a higher level than its surface.

This 'heavy air' seems to be one reason why assimilation does not stop at the colour-drama stage but reaches down into the material process of creating starch. We might express it thus: starch is a bewitched rainbow drawn down into matter by the plant's vital activity.

Starch is subject to many metamorphoses in the plant organism. The most important one is the etherealizing of it into sugar as the sun's warmth draws it upward. Sugar is found in the nectars but is also present higher up, in the still more refined form of glucosides, in the blossom colours. When our 'enchanted rainbow' gleams in a field alight with flowers, it is as though heaven itself were greeting us.

Picturing a plant against the background of the rainbow is not just poetic licence: it is simple realism. And it touches on the deepest questions of the origin of matter.

We often spend years mulling over some idea too vague as yet to find expression. Gradually it takes on contours, becomes clear and transparent, reaches a stage where it can be spoken or written down. Now it is ready for others to examine it. Every artist is familiar with this progression from idea to finished creation. And those who come to know the artist's work resurrect his crystallized thought in their appreciation of it. A culture is nothing more nor less than the realm of human thought made visible.

Must we not think of the creations of the great artist, nature, as works of entirely similar origin? Must not the wealth of forms about us, built by nature according to laws which we are only just beginning to investigate, have issued from a cosmic activity akin to thinking? What happens when a higher than human being thinks, and what effect do such thoughts have?

For those who acknowledge a higher order, questions like these spring from the heart, and one comes to see nature and the universe as the gradual thought-creation of the gods, similar in kind to one's own slowly shaping realizations.

Against the background of such thoughts and feelings the rainbow stands out as the first phase of divine revelation, as the sublime, overarching pattern of fundamental laws that govern all creation – the same laws which, in the realm of life, bring forth the plant. In ancient times the rainbow was felt to be a bridge down which the gods descended. The picture may still serve as a reminder to the inquirer to keep on concerning himself closely with this phenomenon. If one looks into the deeper content of the myth, one comes to understand its message: the creative process travels over the rainbow on its descent into the material world of nature.

Human thoughts crystallized in works of art are resurrected in the soul of the appreciative beholder. Cosmic thoughts embodied in the natural universe come alive again in us when we approach them with such attitudes of soul.

CHAPTER FIVE

The Carbohydrates

When a plant is destroyed, as by heat for example, the remainder consists chiefly of carbon and water. We need only take some plant substance, a leaf or a flower, and heat it in a test tube to see how water condenses on the sides of the tube and carbon lies on the bottom in a form that still shows the original structure. Carbon and water, then, compose the plant's material residues. These are therefore called carbohydrates.

It should be noted that the term applies only to dead matter, for carbon and water can never be combined to make starch or any other plant substance. This distinction between living and dead substance receives too little attention in our time. There is no question here of reviving an obsolete vitalism to which Wöhler dealt a death-blow – perhaps rightly – when he succeeded in synthesizing urea. The nebulous life-force postulated by vitalism could never satisfy a scientific conscience. On the other hand, Wöhler made the mistake of regarding urea as a typical product of organic life, instead of seeing that it is a final, almost inorganic waste-product of life. It will become evident later on that all the chemical elements known to science at the present time are more or less in the same category of waste products – corpses – of organic life.

STARCH

Starch, which is generated by the process of assimilation in the plant's green leaves, is found first in a colloidal, fluid state, in which form it is carried to storage places in the roots and fruit. It is then called starch-reserve, and is found in characteristic grains with a stratified structure. This is the only form in which it can be recovered for industrial purposes. It is obtained by a washing process in which the heavy starch sinks to the bottom of the vessel, while the lighter components, such as chaff, husks and proteins (e.g. wheat germ), are carried off by a gentle stream of water. The dried precipitate is used commercially as granular starch or made into powder for feeding purposes.

Every plant has its own characteristic starch-grain pattern, for the form-principle inherent in a plant comes out in every smallest part of the whole. Going even further, one can say that every part of the world, every landscape, has its typical vegetation; both are formed by the interaction of soil and geographical location with cosmic forces. Looking westward, towards the Americas, we find a distinct trend towards vegetative gigantism. One need think only of the great forest-trees of Canada, of Mexico's giant cacti with the distorted shapes that lend the land-scape its grotesque appearance. The potato is a product of this western world. Its starch comes in the form of giant grains lop-sidedly stratified around a centre out of true, making them look like disfigured goblins. Eastward, towards Asia, landscapes are more ethereal. Palms open their fronds in a volcanic setting, where an outward-raying tendency prevails. Rice is at home in these surroundings. In contrast to the potato, which likes to dwell in darkness, the panicles of rice lift their fruit lightward in loose open sprays. The tiny grains of rice-starch are similarly radiating polygons. They have none of the usual stratification around a centre, thus betraying a tendency to the formless.

So even small starch kernels reveal one-sidedness, and with it the dangers of both West and East. The West is threatened with becoming imprisoned in matter, the East with losing itself in formlessness. Both as to size and shape, the starch granules of wheat, rye and barley show a beautiful harmony between the two polarities. Starches of Central European plants are shaped like miniature suns, even with concentric layers around their cores.

When starches are made into a paste and heated, microscopic examination shows that the granules lose their normal structure. One ring after another explodes, and the matter is absorbed by the surrounding liquid. But this does not resemble solutions of, say, salt or sugar. The starchy fluid is translucent rather than transparent, of a peculiar consistency, neither fluid nor solid, but rather that of paste. This pudding-like condition is called colloidal in chemistry, and the substances that tend to it, colloids. Life-processes take place almost entirely in the colloidal state. We shall have more to say on this subject in a later chapter.

One characteristic of starch is that it turns blue in contact with iodine. This phenomenon is accounted for by its essential nature. We described above the part played by light in creating starch. Now we must explore the characteristics of iodine that account for the above-mentioned reaction. It is well-known that iodine steals light. If a solution of iodine is placed in the path of the spectrum, all the visible part of the spectrum is absorbed. Only the invisible infra-red and ultra-violet rays are left. In combination with light-engendered starch, iodine behaves as it does with the spectrum: it extinguishes light.

We can put this iodine reaction to good use in studying the changes starch undergoes when it is treated with, say, acids. Acid is a dissolving agent, similar to fire. If hydrochloric acid touches the skin, actual burns result. A drop of acid burns a hole through cloth. Acid even dissolves metals. Starch-paste, mixed with a few drops of acid and then heated, undergoes a peculiar change: the colloidal, pasty fluid turns into a bright liquid clear as water. The iodine reaction is no longer blue, but wine-red. Further heating lightens the colour, which changes to red, then to orange, and finally to yellow. Chemical analysis shows that the starch has changed to sugar. This process is applied industrially on a large scale. When the acid is removed and the solution purified and evaporated, starch-sugar, glucose, dextrose, grape sugar and the like are the end-products, available in shops.

SUGAR

The conversion of starch into sugar that takes place in test-tubes repeats a process which the sun carries out with starch in living plants. Like the fiery action of the acid in the test-tube, the sun's heat sublimates plant starches built up in foliage by organic processes and turns them into the sugar stored in blossom nectaries. Sugar, like starch, is a carbohydrate, but a purified one. The chemical formulae for starch and sugar are almost identical. But the molecular weight of starch is many times that of sugar. Starch is denser.

It can be observed that the conversion of starch into sugar by means of acids goes through several stages. Dextrin is an intermediate stage, at which starch has not yet become a sugar but no longer reacts to iodine. This substance is produced commercially by a roasting process. The burning effect of acid is replaced by heating the starch just enough to bring it to the dextrin stage. Bread-crust consists of substances similar to dextrin. The same intermediate states of conversion from starches into sugars appear in the living plant in vegetable dextrins and mucous substances.

Like the starch held in reserve in certain organs of the plant, sugar is stored not only in fruit and blossom, but in the foliage and roots as well. The question arises whether these various sugars are qualitatively the same – whether beet sugar, cane sugar and honey are identical. Or are there certain qualitative differences, especially from the standpoint of nutrition? Rather fierce battles have been fought on this score recently by food reformers and other such groups. We will try to throw some light on the matter by examining the various forms of sugar against a historical background.

From the chemical standpoint, there is little or no difference between beet sugar, cane sugar and honey, or indeed between grape sugar and fructose. Both beet and cane sugar are bioses $(C_{12}H_{22}O_{11})$, while honey, grape sugar and fructose are monoses $(C_6H_{12}O_6)$. The first group has a molecular weight approximately twice that of the second, by chemical standards. Honey

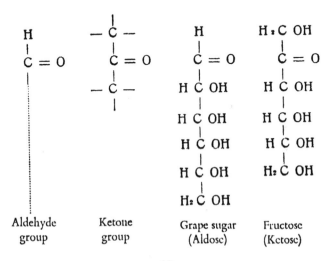

| Aldehyde group | Ketone group | Grape sugar (Aldose) | Fructose (Ketose) |

is a mixture of grape sugar and fructose. From the standpoint of chemical structure they differ, for grape sugar is an aldose and fructose a ketose.

Grape sugar and fructose are distinct also from the standpoint of physics in that solutions made of them behave differently in the presence of polarized light. Grape sugar turns the plane of polarization to the right, fructose to the left, thereby earning the names dextrose and levulose. Fruit sugar is like honey in being a mixture of the two. Beet and cane sugar cannot be distinguished by either chemical or physical methods. Only the above-mentioned capillary dynamic method and biological tests succeed in doing so to some extent, though not conclusively.

Sugar was known long before man sharpened his consciousness of taste with salt. Honey is mentioned in very ancient documents. It was antiquity's only form of sugar. Oriental peoples as well as early Teutonic tribesmen found honey indispensable to their way of living as both a nutrient and an appetizer, as the writings of Plutarch and Aristotle attest.

Alexander's conquests introduced a new factor. He led his armies through Persia to India and there discovered 'a reed that produces honey without the aid of bees'. This was sugar-cane, already under cultivation at that time in India. It was not long before sugar-cane had spread to Persia and Egypt, and it soon became known in all parts of the civilized world. The Arabs discovered the art of crystallizing cane juice. By the eighth century they had a fairly highly developed sugar industry. The sap was boiled down and collected, still hot, in palm leaf cylinders. The finished crystallized cake was history's first sugar loaf.

Charlemagne did much to further spice-trading with the East, which was also the origin of sugar. The Crusades, too, contributed to popularizing sugar in European countries. Columbus introduced sugar-cane to America. That is how Cuba, which devotes a high percentage of its arable land to growing sugar, and San Domingo, came to have such extensive plantations.

Nevertheless, sugar remained more of a luxury than a food throughout the Middle Ages. It was only at the dawn of modern times that the 'enlightened despots', Frederick the Great, Maria Theresa, and Joseph II, decreed that sugar should be made a food and eased its importation by lowering the tariffs and taxes on it. The sugar industry took a tremendous upswing when European refineries began making crystallized white sugar out of the imported brown raw product.

'It was then,' says a contemporary account, 'that further progress was blocked by an artificial product.' In 1800 the German scientist, F. A. Achard, discovered the possibility of extracting sugar from beets. But twenty years passed before a commercially profitable type of beet was bred. Even so, despite all the ingenuity of German technicians and chemists, the industry would not have developed as it did if politics had not played a decisive role.

Napoleon had decreed a continental blockade and thereby stopped all sugar imports. Efforts to produce a beet sugar every bit as good as cane were naturally redoubled. Napoleon himself took a personal interest in speeding the development of the beet sugar industry, with the result that quite a number of factories producing beet sugar were in existence by 1811. After Napoleon's fall, the young industry kept going by virtue of the fact that it had made far greater technical progress than the cane sugar industry. Even today it would be hard to find any other branch of industry where more thought has been spent on working out every detail of manufacture. It began to appear that beet sugar had finally driven cane sugar out of the running, both in the old world and the new.

But again a great war introduced a new factor: the countries involved in World War I used their sugar to make nitroglycerin and other explosives. Their populations, hungry for sweets, had to make do with saccharin, a synthesized by-product of the coal-tar industry. There will be more to say about saccharin in a later chapter. Fortunately, people did not continue using this un-natural sweetener very widely once the war was over.

On the other hand, the war and the disturbed social and economic conditions that preceded and followed it restored cane sugar to a more favourable position. Several factors made for a reduced planting of sugar beets. One was the need for intensive cultivation, combined with increasingly unfavourable working conditions on the farms. Another was the growing exhaustion of the soil due to prolonged monoculture. This favoured insect depredation and disease. The result was a limiting of sugar beet production. Beet sugar could not keep on competing with the cheaper cane product. Indeed, during the decades after World War I, it continued in production only by grace of government support.

The following table gives a graphic picture of these tendencies in world sugar production:

COMPARATIVE PERCENTAGES OF TOTAL WORLD BEET AND
CANE SUGAR PRODUCTION

(from Ullmann's *Encyclopedia of Chemical Technology*)

	1900	1906	1913	1918	1920
Beet sugar	65%	50%	47%	29%	22%
Cane sugar	35%	50%	53%	71%	78%

To return to the question of qualitative differences between the various kinds of sugar, it might be well to consider the facts condensed in the following sketch, bearing in mind that there has been a gradual descent from honey, via cane, to beet sugar – in other words, from blossom, to stalk, to root.

Blossom	Honey	Time of the Patriarchs
Stalk	Cane sugar	{ Alexander, Charlemagne, Frederick the Great
Root	Beet sugar	Napoleon
Coal-tar	Saccharin	Industrialism

Flower, stem and root are the three parts of the plant, clearly differentiated as to form, substance and function. They belong together as three parts of an organism, but each is the carrier of a distinct, essential tendency. For the blossom links the plant with the surrounding macrocosm, the root with the earth, while the stalk (or leaf) is the place where earth and heaven meet in harmonious interaction.

Honey is the immediate product of the cosmic whole. Similarly, life in the time of the patriarchs was permeated and guided from on high by impulses which laid direct hold on the will-life of humanity.

The period of cane sugar's dominance was that of the consolidation of Middle Europe, with its founding of cities and of social orders. It was a time when the heart was especially engaged.

Like the beet, rooted and thoroughly at home in earth, thought is firmly bound to matter in the age of materialism. The world's affairs are guided today by intellects with the hard, down-to-earth, tough quality of rootlike thinking, as yet scarcely touched by the dawn of a new age.

While the extraction of honey requires no technology and the production of cane sugar at first involved only natural tools and objects, such as the moulds made of palm-leaf cylinders, the beet sugar manufacturing process comes from a very highly developed mechanical inventiveness.

A survey of the changing levels of consciousness in human evolution shows how they are reflected in changes in man's nutritional habits. These are again mirrored in the changing use he makes of environmental resources. Sugar is only one of many possible examples.

The history of sugar comes into fresh illumination against this background. Honey, cane sugar, beet sugar, the three representatives of the threefold organism of the plant, take turns as man's companions on his journey through time.

But the pattern of the threefold plant belongs as much to man's organism as it does to his history. Morphological and physiological consideration of the human body show that the same formative forces are at work in the nerve-sense organization of the head as in the plant's root-system. Forces similar to those active in the blossom principle are found in man's will and his metabolism. And the forces that build stalk and foliage work in human feeling, and in the rhythmic system centrally located between head and limbs.

If such connections really exist, then it can be assumed that root substances taken as foods act upon the brain and nervous system; that products of the flowering and fruiting parts of plants affect the metabolism; and that leafy matter stimulates breathing and circulation.

This gives points of view for considering the metamorphoses of sugar as a substance, and hence its effects on the human and social organism in the course of history.

In this connection it can be interesting to observe the use of sugar by the various nationalities. It is certainly noteworthy that the intellectual West has a sugar consumption many times that of the still patriarchal East.

SUGAR CONSUMPTION IN KG. PER PERSON

(from Ullmann's *Encyclopedia of Chemical Technology*)

	1903	1914
England	46·4	40·8
America	32·0	33·6
Switzerland	20·7	34·0
France	20·1	17·7
Germany	19·5	34·1
Holland	17·2	
Austria	10·6	17·0
Russia	6·7	13·3

It should be obvious that we are not here favouring one kind of sugar over another. Our purpose is to give a clear characterization of the various types. Then every reader will know what environmental forces he is dealing with in the sugar he chooses and will be able to make a choice suited to his own individual needs. This approach does away with fanaticism in nutritional matters; for fanaticism flourishes only where a limited horizon shuts out perspective.

CELLULOSE

We showed above how cosmic levitational forces, dominant in the upper parts of plants, work upon the starch of the middle zone and etherealize it into sugar. The densifying earth forces of the opposite polarity harden starchy matter into the cellulose, the wood of roots. Cellulose is chemically very similar to starch, for it is also a carbohydrate, though in hardened form.

Sugar is of course present in all parts of plants, not just in the blossom. Cellulose, too, permeates the whole organism. Though the root contains a far greater proportion, the body of the plant, from root through leaf to blossom, is built of a fine network of cellulose. It comes into the picture as a structural element when plants go beyond the stage of being simply life-processes and begin to take on lasting form and substance. Cellulose is root-related, and the centre of its activity is in the root. Sugar, on the other hand, is the carrier of conversion processes in the plant. It is therefore part of an upward-moving, refining evolution of plant matter from the starchy state into the etherealization of fragrance, colour and pollen.

We described above how starches stored in the plant as a reserve possess definite form-elements, which are no longer present when the starch is etherealized into sugar. The trans-formation of starch into cellulose cannot be observed morpho-logically, but we see its results in the emergence of hard fibrous forms typical of cellulose structure.

Cellulose is so hard that it resists dissolution by biological or chemical means which break down the rest of the plant. It is then mechanically chopped, hammered and combed to remove the waste, and the cellulose comes out as textile fibres. For all their hard, stiff minerality, these still reveal something of the nature of the light-rays which assimilation processes absorbed into the plant; it is as though light had become physically per-ceptible in them. Bundles of shining jute or flax actually look like physical embodiments of the sun's rays.

The purest cellulose, free from any organic admixture, is to be found in seed hairs. That is why the cotton plant is the most productive source of cellulose.

But there is still a way of converting this hard substance to a starch, or even to a sugar. Cellulose shows no iodine reaction, having become so hard and dense an embodiment of light. But if one drops concentrated sulphuric acid on a bunch of cotton, its single fibres dissolve into a formless pulp, and this will show for some time a positive iodine reaction by turning blue. The fire-nature of the acid has etherealized the cellulose to starch. After a while, however, the iodine reaction turns violet, wine-red, orange, and finally disappears.

During World War I, industrial use began to be made of this

process whereby wood can be changed to sugar. Since the product proved capable of fermentation, it was carried a step further and converted into alcohol.

The process for converting cellulose into paste is similar. It is used in manufacturing vellum. Paper is simply a felt made of cellulose fibres. We see them clearly in blotting and filter papers. Making writing-paper requires a coating of glue or some other filler such as satin-white or barite. If water-repellent, parchment-type paper is desired, sulphuric acid must be used to reduce the cellulose to a more paste-like consistency. The felt thereby loses its original fibre-structure and becomes a homogeneous, impermeable, translucent mass.

CHAPTER SIX

Carbon, Hydrogen and Oxygen

Starch, sugar and cellulose are carbohydrates, the most characteristic plant material. All three substances break down into carbon and water when subjected to heat. Now we know that water can be further broken down into hydrogen and oxygen by applying great heat or an electric current. The elements which compose these three substances are therefore carbon, hydrogen and oxygen. This is not, however, a very accurate statement, since carbohydrates, not to mention plants, cannot be built up out of combinations of these 'elements', as the term implies. We have to bear in mind that elements are actually break-down products, or 'corpses' – a description that fits them better in this connection than the term elements.

Nevertheless there are traces of their erstwhile life in these substances and in their chemical and physical behaviour. The following characterizations may perhaps serve to indicate their place in a larger picture.

CARBON

The scaffold-like structure of carbon comes to light whenever organic substance is carbonized. The surface of a piece of charcoal still shows clearly the wood's fibrous composition and other characteristics. Carbon is indeed the form-giving element throughout organic nature. All organic matter leaves a carbon-skeleton behind on its dissolution.

Chemistry has brought the nature of carbon to light in a most remarkable way. Every schoolboy learns that the whole of organic chemistry is built around carbon. Structural chemistry explains this by saying that carbon has the property of being able to combine with itself. Every atom has tentacles. Thus, for example, oxygen has two so-called valencies, or tentacles, which can link up with hydrogen but not with oxygen. So the formula for water is:

$$H—O—H \quad (H_2O)$$

Now carbon has four valencies. A carbon atom in combination with four hydrogen atoms forms a molecule of methyl hydride (methane), thus:

$$\begin{array}{c} H \\ | \\ H-C-H \quad (CH_4) \\ | \\ H \end{array}$$

Carbon has, moreover, the unusual property of being able to combine not only with hydrogen and other atoms but also with itself:

$$\begin{array}{c} H \quad H \quad H \\ | \quad | \quad | \\ H-C-C-C-H \quad (C_3H_8 = \text{Propane}) \\ | \quad | \quad | \\ H \quad H \quad H \end{array}$$

This makes possible a tremendous number of permutations and combinations of molecular structure. Four examples of well-known substances will suffice to show how this unique property of carbon makes for the formation of chains, rings, and all sorts of other ramifications of molecular structure:

Carbon frameworks of

Grape sugar   Camphor   Benzene   Anthracene

We see that carbon does actually form the framework of all these compounds. And the term 'carbon skeleton' is a familiar one in structural chemistry.

It is interesting to compare the number of possible carbon compounds with the number of inorganic compounds that can be made out of the other seventy-two or more elements. Carbon, which combines only with hydrogen, oxygen and nitrogen in addition to combining with itself, is a component of several million known compounds, while the whole realm of inorganic chemical elements – the mineral world – accounts for only a few tens of thousands. This astonishing fact is traceable to the tremendous capacity carbon has for organizing and structuring matter. Whatever conviction one may hold on the score of atoms and structural chemistry, this formative capacity of carbon is the basic reality.

It was no less a man than Kekulé, the founder of structural chemistry, who pointed out with much-needed clarity the dividing line between hypothesis and fact in chemical formulae. He showed that the relative weights of compounds (Avogadro) are based on fact and that the letters used in a formula are its simplest expression. 'But if we ascribe any other meaning to the letters of the formula and regard them as the expression of atoms or of atomic weights, as often happens, the question arises: how big or how heavy are atoms? Since atoms can be neither measured nor weighed, it is clear that hypotheses about their weight are pure speculation.'

We are in full accord with Kekulé in regarding chemical formulae as pictures of rhythmical processes and forces, and, most especially, in looking on formulae relating to the chemistry of carbon compounds as pictures of the formative forces of this substance.

Carbon has a particularly remarkable relationship to iron, a metal possessing centripetal forces which connect it very strongly with the earth. This will become clearer in a later chapter (Chapter Twenty-nine). Since carbon gives physical form to all living matter, it is hardly surprising to find it related to iron.

It is a familiar fact that molten iron can absorb large quantities of carbon, as tea does sugar. But even small amounts of carbon change the nature of iron, making soft malleable iron into brittle cast-iron or tensile steel, depending on the amount of carbon and the cooling method used. It is noteworthy that here again carbon functions as a hardening and form-determining agent.

On the other hand, iron changes carbon into diamonds under

certain conditions. If carbon is dissolved in white-hot iron and the solution subjected to sudden cooling, the carbon is precipitated in the form of tiny diamonds. Diamonds are pure carbon, the hardest as well as the most shining of earth's substances.

Carbon, the shaper, is strangely enough present also in air as carbon dioxide. It is air's heaviest component. This causes it to seek out low places. As we have seen in the case of the Dog's Grotto, it sometimes covers the surface of the ground like a lake.

Plants absorb carbon-dioxide and use the carbon to build up form. For carbon is always the element which brings fixed form into living processes. For this reason it could truly be called earth-substance instead of carbon.

HYDROGEN

Hydrogen, in contrast to carbon, is the earth's lightest substance, hence given to seeking out air's highest levels. The lower levels of the atmosphere contain almost no hydrogen, usually only 0.2%. Investigations of the stratosphere indicate that hydrogen increases with increasing height. At 100 miles (approximately 150 kilometres) hydrogen is said to measure 99.5%. This tendency of hydrogen to rise has been put to use in balloons and zeppelins.

Hydrogen's chemical properties reveal the same tendency. In combination with hydrogen even heavy substances such as lead change into gases. The following examples serve as illustration:

Marsh gas (methane), a hydrocarbon	CH_4
Will o' the wisp, phosphene	PH_3
Rotten egg odour, sulphide of hydrogen	SH_2

Hydrogen is also connected with warmth in ways that merit attention. It has the hottest of all flames. Iron and steel are welded with an oxyhydrogen torch that uses a mixture of hydrogen and oxygen, and hydrogen is the source of heat in all other autogenic welding processes. Hydrogenous processes are invariably accompanied by heat formation.

Zinc dissolved in hydrochloric acid gives off bubbles of hydrogen. The chemical formula is:

$$Zn + 2HCl \rightarrow ZnCl_2 + H_2$$

But if we substitute sulphuric for hydrochloric acid, the same thing happens: zinc disappears and hydrogen bubbles rise. The formula here is:

$$Zn + H_2SO_4 \rightarrow ZnSO_4 + H_2$$

Any acid can be substituted for hydrochloric acid, with the same effect as far as hydrogen's part in the reaction is concerned. For hydrogen is the element common to all acids; active hydrogen, that is, known to chemistry as ionic hydrogen, which gives acids their essential character. The above formulae might suggest that there is a primary affinity between zinc and chlorine, or between zinc and the remnant of sulphuric acid, with the release of hydrogen only a secondary phenomenon. The findings of the ionic theory contradict this view.

After what we have seen of the qualities of hydrogen, we can say that it is hydrogen, with its tendency to rise and to liberate warmth, that is responsible for the dissolution of metals.

Now the question arises whether this tendency is to be regarded as a purely physical phenomenon of anti-gravity, or as the last visible remnant of a cosmic fire-force that pervades the universe as a dissolving, de-materializing element?

We have already described the role warmth plays in the process whereby plant substance is etherealized from the middle zone upward: the process whereby starch changes into sugar and sugar into still subtler substances, as can be observed when summer's heat causes the plant to pour itself out into the universe as fragrance, colour, pollen-dust.

The materializing force of spring conjures into visibility all the wealth of greenery and foliage that marks its climax. Then summer takes over, and with its sublimating warmth brings forth the ethereal marvel of the blossom. Flowers are that portion of the plant where substance is refined and emanated. Scent and pollen rise sunward and stream out into endless reaches of the universe. In Goethe's sense, the plant has a real, enduring being. With its material decay, its essence or 'idea' returns to the cosmos, leaving the tiny, largely mineral seed as an anchor, a guarantee that it will reappear on earth when conditions warrant. Investigations of the stratosphere are credited with discovering clouds of

pollen, still mounting skyward, many miles above the earth.
If this is true, must not hydrogen be somehow intimately involved? It has been found that hydrogen processes predominate
increasingly as the time for blossoming approaches. The plant-
matter that has become so light and airy now ascends into the
heights as it were on wings of hydrogen. This will be more fully
described in later chapters.

Hydrogen means, literally, water substance. But the word has
a false ring to it. Where did it come from? Lavoisier, in 1783,
produced a gas from water and gave it the name hydrogen
because of this. Cavendish and Watt discovered that burning
hydrogen gives off water, a fact that fixed the name in use. A
more penetrating study of this element yields very slight evidence
of a relationship to water. Water actually contains only a small
amount of hydrogen: 11%. But even in this dilution hydrogen
imparts something of its buoyancy and warm, releasing character
to water, making the latter a dissolving agent for salt, sugar, and
other substances.

If hydrogen were to be baptized with a name indicative of its
inner nature, we would have to call it 'pyrogen' (fire-substance).

The physiological processes accompanying the spiritual fire
called enthusiasm are also hydrogen-based. A fire-force works
in our enthusiasm which, radiating from the heart, warms our
entire being. It actually makes us feel as though our blood were
lighter. The genius of language teaches us to say that we are
'fired', 'hot-blooded', and the like. Enthusiasm lifts us above our
difficulties, our everyday worries, and on occasion above some
of our material limitations. The heart is the centre of this process,
which has power to overcome even physiological problems.
How clear the medical logic in the use of hydrogen-rich oils
and seeds becomes, when one considers the strength of the fire-
process in them! Despite differences in application, they have a
common capacity to 'fire' the bodily functions and thus to
heighten the mastery of soul over body.

OXYGEN

When one is asked where oxygen is found, one tends to
answer, 'in the air'. The longing for oxygen is in our blood, and

we think we derive it from the air in breathing. But oxygen constitutes only 20% of the atmosphere. We shall show later that the remaining 80% is at least as important. Our longing for oxygen is really a longing for life, which oxygen supports. But is this carrier of life to be found only in the air, where it is present in such small amounts? When we look at the great expanses of water on our planet, at the lakes and oceans, streams and rivers, we are looking at a gigantic oxygen reservoir, for water is 89% oxygen. This is the source of life for the whole earth's vegetation. The fructifying rain, the life-giving rivers, owe their enlivening properties to oxygen. Without water there could be no life on earth. Plants, which consist principally of water, might even be called organized water, hence organized life. Man and animal share as well in the universal life of which water – or the oxygen active in it – is the carrier.

But what is the nature of this life?

What we mean here is simply and solely the vital process involved in plant growth. Oxygen is its carrier. A new-born baby becomes a citizen of earth with his first independent breath of oxygen. Oxygen 'enlivens' every earthly organism. It is the expression of that life-force which presses into material embodiment when, in burgeoning spring, the sap rises and all manner of leaves, grasses and weeds burst forth.

Oxygen is in this sense the antithesis of hydrogen, which we came to recognize as the bearer of the 'being' or 'idea' of plants, as the element which carries them out into the cosmos on its wings. Oxygen brings about just the opposite occurrence. Borrowing Goethe's terms, we might describe oxygen as the bearer of forces whereby 'being' becomes 'appearance'. When nature stirs again in spring, the idea of the plant begins incarnating, reaching a peak of embodiment at midsummer when it blossoms and sets its fruit. It has then spent itself as to appearance. As it fades away and withers, leaving nothing but seeds, the being of the plant withdraws again from manifestation. In the following growing season it uses the seed as an anchor-hold for a further reappearance.

This marvellous rhythm of being and appearance, of blossoming and germinating, of expanding and contracting, is the primal phenomenon described in Goethe's doctrine of metamorphosis. Though he did not state it specifically, the metamorphoses to which he called attention in the expanding and contracting rhythm governing leaf formation are smaller cycles of the cosmic rhythm holding sway in the alternation of being and appearance.

This rhythm, which uses the activity of fire-force (hydrogen) and oxygen as tools, also comes to light in certain phenomena involving dyes and leuco compounds. Dyestuffs can be made to disappear in the test-tube with the aid of hydrogen (hydrochloric stannous chloride solution → nascent hydrogen). The solution becomes colourless. What has happened? Two possibilities exist. Either the colour was destroyed, as occurs when chlorine is used for bleaching, or the dye-colour was raised to a plane where it became invisible. But if it is true that hydrogen caused spiritualization to take place, then it should be possible to conjure 'the colour back into appearance with oxygen reactions, such as can be induced with hydrogen peroxide. And a drop of the latter actually brings this off. Dye chemists speak here of a leuco compound which is colourless and yet can be made from dyestuffs with the aid of a reducing agent. Oxidation makes the colour reappear in the compound. Technical use is made of this process in vat-dyeing. It is impossible, for example, to dye directly with indigo, because of its insolubility. An indigo vat is therefore first set up and the indigo made into a colourless solution by reducing agents. The cloth to be dyed is then soaked in this colourless indigo solution and hung up in a draughty place. Oxygen causes the originally colourless material gradually to turn blue.

It is in character for oxygen, as the carrier of life, to bring movement (i.e. chemical activity) into even the mineral kingdom. It combines with almost all substances and makes them capable of chemical reaction. Silicon, calcium and other elements become chemically active only when they have combined with oxygen, which enables them to become silicates, lime, and so on.

What is the origin of the German name for oxygen – *Sauerstoff*, 'sour substance'? It was found that certain substances with an alkaline reaction take on an acid character under intensive oxidation. Manganese oxide, for example, is a metallic base, and so combines with sulphuric acid to form manganese sulphate:

$$MnO + H_2SO_4 = MnSO_4 + H_2O$$

However, manganese heptoxide is an acid (or, strictly speaking, an acid-forming agent). It therefore combines with alkalis such as caustic potash to form permanganate of potash:

$$Mn_2O_7 + 2KOH = K_2Mn_2O_8 + H_2O$$

Apart from the fact that this phenomenon is limited in practice to metals of the iron group, some of the strongest acids, such as hydrochloric, contain no oxygen. Considering how superficial and limited oxygen's acid-forming capacity is, it is surprising that it was given a name so uncharacteristic as 'sour stuff'. But what would be the fitting term? Since it accounts for almost 90% of water and is so active in and through this element, the name water-stuff (applied to hydrogen in German) would be more appropriate. However, as this would cause confusion we will settle for 'biogen' (life-substance), an acceptable designation considering that here on earth life and water are so inseparable.

The rhythmical interaction between 'pyrogen' (hydrogen) and 'biogen' (oxygen), being and appearance, expansion and contraction, seems stabilized at every stage in carbohydrates by the earth-substance, carbon. If carbon were to function unrestrictedly, the rhythm would cease and the carbohydrates take on a rigidity such as we have noted in cellulose, the plant's woody component. If no bounds were set to pyrogen (hydrogen) the carbohydrates would become formless, as they do in sugar, colour, scent and pollen, and would be etherealized away into the cosmos.

When we describe carbohydrates as the product of interacting carbon, pyrogen and biogen, we are not speaking of an atomistic combination of these three substances, but rather of an interweaving of cosmic qualities which produce a unified substance, starch, in its various metamorphoses. It is only when this organic unit is destroyed that the three substances fall away like 'corpses' from the living organism. In speaking as we do above of carbon, oxygen and hydrogen, we are referring, therefore, to the spiritual archetypes of these substances: to cosmic formative forces, cosmic life forces, cosmic fire forces. It is certainly possible to conceive of earthly substances as material forms of cosmic realities and to see in them 'incarnated ideas', like the idea of the archetypal plant incarnated in the many forms of earthly vegetation.

A dynamic-chemical symbol of carbohydrates might then conceivably look thus:

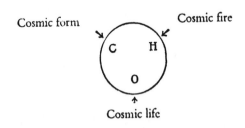

Cosmic form   Cosmic fire

C   H

O

Cosmic life

Life oscillates between form and fire, rigidity and dissolution. When life moves to the etherealizing pole, then water, the 'solvent of life' (H_2O), is its expression. When it becomes tied at the opposite pole to carbon, 'earth substance', the carbonic acid (CO_2) produced thereby may be termed 'paralysed life'. Chemical formulae take on a new reality when they are extended in this way.

From this standpoint, it would be appropriate to call carbohydrates 'formed and fire-quickened cosmic life'.

It can certainly be assumed that the three cosmic qualities which interact to form carbohydrate vary in intensity in its three metamorphoses, starch, sugar, and cellulose. This is by no means a new conception; it is suggested by the allotropic modifications of a great variety of inorganic substances. Yellow phosphorus, for example, is active, burns on exposure to air, is extremely poisonous and a source of light. Red phosphorus is inactive, non-poisonous and so to speak dead. What can be seen here in an inorganic substance could surely obtain to an even higher degree in the fluctuating interaction of entities not yet materialized. Starch may therefore be termed a carbohydrate in which life predominates, sugar a carbohydrate in which pyrogen has the upper hand, cellulose a product of cosmic shaping forces. These three metamorphoses may be symbolized as follows:

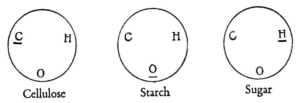

Cellulose   Starch   Sugar

Modern chemistry sees cellulose as a product of the proliferation of starch molecules through polymerization. This is again an expression of the primary reality of cosmic shaping forces working in material condensations.

CHAPTER SEVEN

Scents (Etheric Oils) and Resins

The sugar produced from starch when hydrogen acts upon it has a tendency to go further in the same direction, towards etherealization. If sugar solutions such as grape juice are allowed to stand, they ferment, and the sugar separates into carbonic acid and alcohol. A penetrating study of these fermentation products would say that on the one hand biogen (oxygen) is tied to carbon as carbonic acid, while the hydrogen is greatly reinforced by the alcohol. This alcohol is a strange substance. It is very volatile, aromatic and inflammable – all indicative of the predominance of cosmic fire. Some peoples call alcohol 'firewater', with good reason. It is also a familiar fact that alcoholic fermentation gives rise to all sorts of aromatic compounds known to the connoisseur of wines as their 'bouquet' (in German, 'flower').

The emergence of this bouquet in wine-casks stored in subterranean vaults is a caricature of the wonderful process that occurs in every blossom growing between earth and sky. Flower scents, known to chemists as etheric or ethereal oils, all contain a great deal of pyrogen (hydrogen). They are therefore obviously similar in nature to this substance. This is shown in their great volatility; they fly away, as it were on wings of pyrogen. Moreover, they are highly inflammable, as everyone knows who has made the pleasurable experiment of squirting the etheric oils of orange peel or Christmas tree twigs into a candle flame and has stood amazed at the splendid fireworks that result.

It is also a familiar experience to find these aromatic substances becoming the more fragrant the more they follow the law of their own being: etherealization. Concentrated oil of roses, lavender or fir smells quite unpleasant; it takes a certain amount of thinning for the scent to emerge. The perfumer's art lies in knowing just where to stop the thinning process.

How are these oils extracted? There are two methods: steam distilling and enfleurage. Bulgarian attar of roses is extracted by steam distillation. Rose petals are mashed to a pulp in water and then brought to a boil in stills. The rising steam carries off the fragrance of the petals. In cooling, the rose oil collects on the surface of the water condensed from the steam and is drawn off. This is the method used in extracting oils of rosemary, lavender and evergreens. These etheric oils have only a few external qualities in common with the fatty oils we shall discuss later: they float on water, cause transitory grease-spots to appear on paper, and the like.

More delicate scents such as lilac, jasmine and verbena are extracted by a more complex process. Sheets of glass are smeared with butter, which has been allowed to stand so long – sometimes for a century or more – that it has come to look like glycerine or jelly. The blossoms are laid on these buttered surfaces, which take up the scent. Butter prepared in this way greedily absorbs the etheric oils. The scent-saturated butter is then distilled, the perfume being extracted by double distillation, while the butter is used again in further manufacturing. This process is called enfleurage. The aged butter is a greatly valued asset of perfume factories, almost worth its weight in gold.

Now there is an obstacle to the plant's etherealization, to its streaming out into the cosmos. This is the forming of resins, a condensation process which is the polar opposite of sublimation. Biogen (oxygen) is its carrier. A comparative study of the chemistry of resins and etheric oils shows that though the two groups are very similar, resins contain considerable oxygen.

Pine needle oil (main component: pinene)

$C_{10}H_{16} = H_2C$ ⟨ $CH_3 - C - CH_3$ ⟩ CH

$C_{20}H_{32}$

$C_{20}H_{30}O_2$ — Pine resin (pimaric acid)

Anyone who has stored etheric oils will have found that as time goes on the bottle-stopper gets stickier and stickier and a resinous crust finally forms, while the resin odour grows more pungent. The process can be greatly accelerated by exposing these oils to the air. We see here how oxygen makes the rising stream of scent precipitate as resin.

The milky juice of certain other plants, such as dandelion, spurge and rubber, belongs to this same group of resins.

Evergreen resins are extracted by cutting the bark and catching the sap in cups. Resin thus extracted and purified is known commercially as turpentine. So-called Venetian Turpentine is a product of larch resin.

More valuable products are obtained through distillation. The first, volatile components to be separated out when turpentine is heated are spirits of turpentine, a substance closely related to etheric oils. The next in line is oil of turpentine. The sediment is colophonium. Other resins also yield 'spirits', oil and sediment. Even fossil-resins such as amber yield the medically prized spirits of amber and oil of amber.

This way of looking at resin-formation can lead one to feel that the plant sacrifices something when it lets some of its substance stay behind on earth. The part this sacrifice plays in seed formation will be discussed later. Perhaps this is one reason why antiquity, having some inkling of these matters, valued and used resins in sacrificial rites. When incense (made from the resin of *Boswellia carteri*) burned on the alter, the priests of those days may well have felt the harmony between heaven and earth in the ascending smoke of the sacrifice thus:

Biogen has rained fragrance down as resin. Now cosmic fire resurrects the resin as a fragrance.

CHAPTER EIGHT

The Fatty Oils

It is a striking fact that oils are almost entirely restricted to one plant organ, the seed. In our latitudes, crops such as rape, hemp, flax, poppy and sunflower are grown for the oil-content of their seeds. Overseas, and in more southerly climes, olives, cotton and peanuts are among the plants grown for their oil content, found in fruit and seed.

We may ask where this oil comes from. Its lineage is scarcely traceable to starch or sugar, at least not from the chemist's standpoint. But we note that oil is formed in ripening seed – at a time when the visible plant has passed its zenith and is pouring its substance out in scent and pollen, and when the seeds are then bathed in summer warmth. One may then be inclined to think that what is seen here may be a further example of the plant's many cycles of expansion and contraction – in this case a rhythm of condensation whereby the material seed is prepared for overwintering.

We described above how scent uses wings of pyrogen (hydrogen) to carry the plant being out into universal space, and how, in the following spring, this being seeks out the anchoring seed to re-manifest. This phenomenon takes place within the framework of the law of metamorphosis. The polar rhythm, expansion and contraction, being and appearance is its pattern. But what for the plant being is expansion and contraction is just the opposite on the plane of physical appearance. Expansion, for the spirit, is material contraction, and vice versa. Thus, when the plant expands into the cosmos, its physical manifestation shrinks into the compass of the seed. The oil that forms in seeds under the influence of summer's heat is an inverse reflection of the plant's outstreaming. Cosmic fire in the sun's warm rays is condensed into oil, as it were concentrating forces of physical expansion in readiness for next spring's germination.

Chemical analysis of oil brings to light the huge proportion of pyrogen (hydrogen) present in it. This can be read in the following formulae:

49

Olive oil's chief component, trioline: $\qquad$ $C_3H_5(O.C_{18}H_{33}O)_3$

Rape seed oil's chief component, trierucine: $C_3H_5(O.C_{22}H_{41}O)_3$

But it is also a fact that oil can be split into two components when broken down by alkalis, for example. One of the components that emerges is a fatty acid; the other, alcohol-like glycerine. Are these breakdown products possibly also the result of a cosmic-biological process?

We have previously described how fermented sugar produces alcohol and carbonic acid, thus dividing and tending in two directions: toward a concentration of pyrogen on one hand, of biogen on the other. If something similar occurs in the blossom with the formation of etheric oils, which are breathed out and disappear, then we should expect to find somewhere else in the plant a complementary acid rich in oxygen. And we do find it in the swelling seed-bud. We might perhaps put it this way: the plant acids in the developing seeds are like a vessel in which the scent flowing back from the periphery of the cosmos is gathered up. The fire-force in this essence warms and permeates the developing seed-organisms. Oil is the product of this interaction. In this sense we might call oil 'the perfect plant substance'. We began with starch, 'the virgin substance', and progressed by way of sugar and fragrance to the final cosmic-biological synthesis in the perfected substance, oil.

The ancients must have known instinctively about this process, for they used oil to anoint the kings and priests who in their dignity were to represent perfection, the highest unity.

Cold-pressed oils from seeds and fruits yield a wonderfully clear, almost transparent oily fluid. This makes ideal edible oil, known to technology as 'virgin oil'. Second and third pressings, aimed at extracting every drop, are made at constantly increasing temperatures. Their yields range in colour from yellow to brown. They too are used as food-oils, after being refined and bleached. The cakes left after the last drop is extracted are treated with fat-solvents such as benzine or carbon tetrachloride and made to yield still another oil, used in manufacturing.

In addition to edible oils, made chiefly from olives, rape-seed or peanuts, there are oils that have a tendency to form resin on exposure to the atmosphere. As we saw above, etheric oils, too, have this tendency owing to the effect of atmospheric oxygen. We showed how plants prepare to form fatty oils by first pouring out etheric oils in their wafting fragrances. Resin formation might be looked upon as a remnant of oil's childhood. Resinous oils are called drying oils because of their tendency to form dry, resinous surfaces on contact with the air. This makes them suitable for use in paint. Flax, poppy and sunflower are the main sources.

The fire-relatedness of oil was put to practical use in the oil-fuels of former times. Rape-seed oil was the principal source.

More will be said later on the subject of animal oils and fats such as butter, tallow, lard and whale-oil. These are similar to plant oils, but harder. Man's ever growing need for these denser oils and fats has developed processes whereby the normal consistency of oils is artificially changed to that of lard or butter. The raw materials used in this fat-hardening process are the plant oils or whale-oil, and its product is margarine.

When animal fats are broken down or split, they yield, in addition to glycerine, a fatty acid quite similar to that derived from vegetable oils, except that it is harder and will even crystallize. Stearine, obtained from beef tallow, is an example.

The alkaline salts of these fatty acids are soaps. Potash soaps have a salve-like consistency and are known as soft soaps, while sodium soaps are harder and can therefore be made into cakes.

Oils and fats boiled up with an alkali form a soapy mass containing glycerine as well as pure soap. This crude soap is called glycerine soap for that reason. Curd soap can be separated out of this solution by adding salt. Common salt is used here for its coagulating effect. It separates solids from liquids in colloidal solutions, of which soap is one. When the solid soap has been removed, the glycerine is extracted by distillation from the lye.

All soaps can be decomposed by any acid stronger than fatty acids, such as hydrochloric, sulphuric, or even acetic acid, for example. When a soap solution is decomposed with acid, the oily or fatty acid separated off rises to the surface of the fluid. Depending on whether the raw material used in the soap was a vegetable oil or an animal fat, the separated product is either an oily liquid known as olein or a solid crystalline block such as stearine, for example.

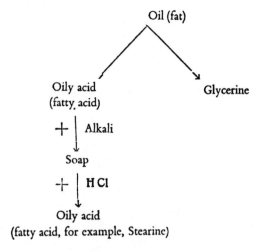

Modern methods of splitting fats no longer always take the roundabout path of soap making. They follow the direct one that employs catalysts, enzymes, and super-heated steam.

CHAPTER NINE

Protein

The body-building material, as typical in animals as carbohydrates are in plants, is protein. But the plant, too, manufactures protein, particularly at one stage of development, that of seed formation. This is where the animal and plant worlds meet.

At the stage where the plant is exhaling scent, the life active in its blossoms is a strong attraction. Butterflies come winging to the out-poured fragrance. Bees arrive to drink the nectar. Beetles are drawn to the scent and linger happily around it. Bees, butterflies and beetles come and go, visit other flowers, thus carrying on the fertilizing process. A constant rhythm of approach and withdrawal, enticement and fructifying takes place here as part of the larger rhythm of seed formation and etherealization.

The movement and rhythm of the insect world find expression in the seed substance known as protein. This is not to say that animal substance is transferred directly to the seed by insects; it is rather the active exchange between plant and universe taking place in all this movement which finds physical expression in seed-protein. It is of no consequence whether this fertilizing exchange is effected with the help of insects or of the whole atmosphere, as in the case of wind-pollenization.

When wheat flour is made into dough and kneaded in a stream of running water until the starch is gradually washed away, a sticky, stringy mass is left. This no longer contains starch or any other carbohydrate, but only a protein called gluten. Protein in this form is found in seeds of every kind.

We described above how a 'carbon skeleton' is left when carbohydrates are destroyed by heat. Protein also leaves a carbon remnant when it is burnt, but the process is livelier, with foaming, bubbling and an exuding of unpleasant odours. Who does not remember the smell of burning hair or milk – both of them proteins? Animal protein (in hair, milk, meat, wool, silk, glue, etc.) behaves like vegetable protein. Since the typical plant substance is carbohydrate and the typical animal substance is protein, the protein found in seeds attests to plant and animal worlds having met there.

We can discover something of the nature of proteins by breaking them down. Caustic soda can be used, as in processing oil. After prolonged boiling the mixture gives off a penetrating odour recognizable as ammonia (NH_3), a nitrogen compound. Here we discover traces of a basic substance not yet encountered in the vegetable kingdom.

When protein is broken down with potassium and the breakdown product is dissolved in water and treated with iron salts, a beautiful intense blue called Prussian blue appears in the solution. It is interesting to follow what has happened here. Potassium, a substance with a strong affinity to life, has withdrawn hydrogen and oxygen, the elements present in water, from the protein. Carbon and nitrogen, the components of cyanide, are left. This substance is one of the deadliest poisons known to man. But potassium cyanide in combination with iron salts produces Prussian blue, which is not poisonous.

$6KCN$	$+ FeSO_4$	$\rightarrow K_4Fe(CN)_6$	$+ K_2SO_4$
Cyanide of	+ Ferrous	→ Potassium	+ Potassium
potassium	sulphate	ferrocyanide	sulphate
$3K_4FeCN6$	$+ 2Fe_2(SO_4)_3$	$\rightarrow Fe_4[(Fe(CN)_6)]_3$	$+ 6K_2SO_4$
Potassium	Ferric	Prussian	+ Potassium
ferrocyanide	sulphate	blue	sulphate

We may say that iron 'heals' the poison by making it non-poisonous. And were it not for iron's healing property we should be constantly poisoned by the cyanide compounds formed in the process of digestion. The iron in our blood, however, instantly transforms these compounds into harmless ones. (Cf. Chapter Twenty-nine.)

A more exact analysis of protein shows it to be composed of four elements: carbon, hydrogen, oxygen and nitrogen. Its gross formula is given as:

$$C_{720}H_{1134}N_{218}O_{248}$$

But the structure of this protein molecule is anything but clear. We know that it can be split into smaller complexes known as peptones by careful disintegration, such as it is subjected to by enzymes present in saliva and the stomach (pepsin). These in turn can be split into still smaller complexes known as peptides under careful disintegrative action, for example by trypsin, the intestinal fermenting agent. These can with care be broken down still further by pancreatic enzymes, for example, or by being carefully boiled with acids and alkalis. This separates them into the so-called building blocks of protein, the amino acids.

Now all this is very interesting, but protein becomes unimaginably complicated. Does it not make us suspect that protein is only so complicated when analysed, or – in other words – broken down? For all analysing is actually destruction. Is it not conceivable that living protein, as it comes from nature, may not really be at all complicated?

For the time being let us keep in mind just the fact that protein is the product of four interacting elements: the formative, life and fire forces we already know, plus a fourth entity which finds expression in nitrogen.

CHAPTER TEN

Nitrogen

W here is nitrogen to be found in nature? Practically speaking, it is present in a free state only in the atmosphere. In fact, from a quantitative standpoint it is the carrier of air, for air is 80% nitrogen to 20% oxygen. Let us now consider whether nitrogen is suited to be air's carrier in other, more essential respects.

A study of the breathing process shows that the nitrogen inhaled is exhaled again, laden with moisture and carbonic acid. There is no change either in its composition or in its volume. What purpose does nitrogen serve, then, if it comes out exactly as it went in?

When one tries to live into the movement of nitrogen streaming in and out in breathing, it can be experienced as an oscillating exchange between man and world; nitrogen moves in pendulum-like rhythm back and forth between the two. So we come to regard it as the carrier of motion and rhythm, enabling oxygen to be inhaled, and the used air to be expelled again.

What would happen if there were no nitrogen in the atmosphere? We should be burnt up by the concentrated oxygen. The dilution of oxygen by nitrogen makes breathing and its rhythm possible. Even the slightest variation in the nitrogen content of the air around us quickens or slows breathing to a surprising degree. Medical science has come to realize how great a role rhythmic breathing plays in health and illness.

A person in normal health breathes eighteen times per minute. This adds up to 1,080 breaths per hour and 25,920 breaths per day. The latter number has, as we shall see, a cosmic reference.

In the course of a year, the sun makes one complete round of the zodiac. It rises on March 21 at the vernal point, which today is in the sign of Pisces. Day by day its rising-point progresses, from Pisces to Aries, to Taurus, to Gemini, and so on, returning after a year to the sign of Pisces. This vernal point, however, is not identical with that of the previous year, but slightly behind it. As millennia go by, the sun's vernal point thus travels around the Zodiac in a direction counter to its annual course. At the time of Christ's birth, the sun rose on March 21 in the sign of Aries. Thus, in the approximately 2,000 years that have elapsed since then, the vernal point has travelled through almost one entire constellation. It takes 25,920 years for the vernal point to go all round the zodiac. This period of time is known as a Platonic year.

We see, then, how our breathing keeps pace with the great rhythm of the sun. One human breath mirrors a solar year; one day's breaths reflect the number of solar years in a cosmic year.

This throws light on nitrogen's role as carrier of rhythm and movement, as well as of the airy element; nitrogen repeats universal rhythms in the microcosm, man.

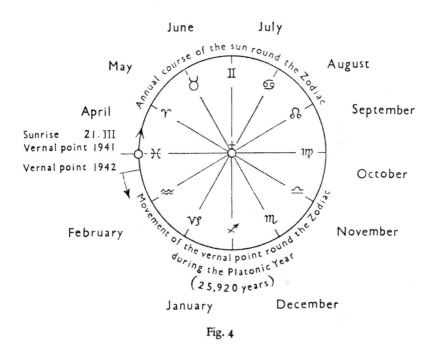

Fig. 4

A still clearer picture emerges from a study of the movement of air currents around the earth. The air mantle is in constant rhythmic motion. Trade winds, monsoons and other well-known air currents are not the only phenomena of this rhythmic patterning: the atmosphere as a whole is a manifold, rhythmically moving organism which serves as the earth's breathing system. Nitrogen is the carrier of this breathing as it manifests in wind and storms and weather.

Again, nitrogen seems inappropriately named. It should be called movement-substance, or air-stuff.

Nitrogen is of such a nature as to lend itself to being a carrier of feeling as well as of breathing. Everyone knows how closely related to feeling breathing is. When we feel joyful our breath quickens. When grief weighs on the soul, breath comes slowly and heavily. Sanguine people have a faster pulse and breath-count than melancholic natures do. Breathing is a constant rhythmic mediating between man and his surroundings. Every breath we draw brings the outside world into us. This enables us to have a feeling of our surroundings, similar to a touching with hands and fingers or a grasping with thought. One of the greatest achievements of the new approach to man inaugurated by Rudolf Steiner was the discovery that the breathing system is the physiological basis of feeling, as the nervous system is of thinking.

A distinctive social fact is that we all breathe the same air; there is nothing we do so much in common as breathing. All other possessions tend to be individually owned, and people even go out of their way not to share objects of personal use. No modern person likes to eat out of the same bowl with others. But we all enjoy the air in common. Certainly there are fresh-air fiends who even want their own air to breathe and cannot bear to be shut up with others in a single room. This is especially true of the English. And what is the reason for it, if not egoism? Is the fact of my inhaling something of another person's being not a way of feeling what his nature is – feeling it lovingly?

Since carbohydrate is the substance of plants, they are bound to the soil and have the power neither of movement nor of feeling. The typical animal substance is protein. This is in part the product of cosmic movement, which comes to physical manifestation in nitrogen. And in the blossom, where plant and animal worlds touch and commingle, the plant-protein of the seed grows as the fruit of that meeting. In seeds, with their freedom to separate from the mother plant and seek new homes, plants acquire a certain mobility.

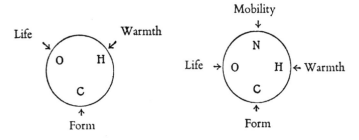

Carbohydrate (plant substance)   Protein (animal-substance)

It is significant that nitrogen is found only in the free state in inorganic nature, and never as part of any chemical compound. Since nitrogen is the carrier of movement, it cannot be bound. It must be free to move. But technology has succeeded with enormous effort in tying nitrogen to oxygen. The two substances are forced under pressure through an arc of electric flame (in the Birkeland-Eyde-Schönherr process). This yields saltpetre, or nitric acid, a material used in the manufacture of nitro-cellulose (gun cotton), nitro-glycerin, picric acid, and all the other modern explosives. Saltpetre was even an ingredient of the old-fashioned gunpowder invented by the monk Schwarz.

Now what is an explosive if not imprisoned motion? And there is, in fact, scarcely a single explosive that does not contain imprisoned nitrogen.

CHAPTER ELEVEN

The Cosmic Nature of Earth's Substance

It is natural to ask whether the forces which work in nature, and whose imprint we find in nature, could not be traced back into the cosmos. In other words, what is the cosmic origin of those forces which we have called cosmic fire, life, form and movement, and which express themselves physically in hydrogen, oxygen, carbon and nitrogen? And are most of earth's other familiar substances also an expression of this cosmic interweaving, and to be understood in relation to it?

If the chemical elements are arranged in a sequence based on a scale of increasing atomic weight, their characteristic qualities reappear at definite intervals. We find that the seven elements in the first series, arranged according to weight, characteristically show no interrelationship:

Series 1.	Li	Be	B	C	N	O	F
	7	9	11	12	14	16	19

But if we proceed beyond fluorine, the next higher atomic weight belongs to sodium, and we see at once that it repeats lithium's essential qualities. Magnesium, which follows it in the second series, shows a like relationship to beryllium, aluminium to boron, silicon to carbon, phosphorus to nitrogen, sulphur to oxygen, and chlorine to fluorine:

Series 2.	Na	Mg	Al	Si	P	S	Cl
	23	24	27	28	31	32	35·5

The next substance in the weight-series is potassium, with qualities which are at once recognized to be almost identical with those of sodium and lithium. This is the start of a third octave, with potassium related to sodium, calcium to magnesium, scandium to aluminium, and so on:

Series 3.	K	Ca	Sc	etc.
	39	40	44	

If we proceed through the whole list, we find twelve such octaves. Details and exceptions will be discussed later. The significant fact discovered in the series is that matter is subject to a rhythmic ordering. Even atomic weight is the expression of rhythm, as we showed above. It is really not surprising that the law of single and multiple proportion (Avogadro), which defines the rhythmic character of matter, reappears in essence in a new metamorphosis as Newlands' law of octaves. This law was later developed by Meyer and Mendelieff into the Periodic Table of the Elements.

Just as the laws discovered by Avogadro were narrowed down into a spatial picture of the atom, so the rhythmical principles expressed in the Periodic Table were developed into a spatial concept which accounted for the rhythmical qualities of substances by relating them to the number of electrons attached to an atomic nucleus. The search for a qualitative explanation of the varieties of matter seemed to call for the existence of a kind of primal nucleus, surrounded by electrons in various numbers and orbits, from which the qualitative periodicity of substances emerged (Moseley).

We are not required to oppose these hypotheses, so long as we remain aware of the reality that lies behind them. Goethe did not battle against Newton's wave theory of light in itself, but against the belief that it explained the reality of colour. The balancing of polarities always issues in rhythms. We can therefore expect to find a rhythmical quality – expressed in the wave-character of colours – as the balance between light and darkness. But colour-vibrations are to be understood as merely the physical manifestation of the nature of colour. They are no more the reality of colour than an anatomical description of a human body is the man himself. Similarly, hypothetical pictures of the nature of substance, developed in theorizing about atoms and electrons, are to be considered simply as mathematical expressions of physical aspects of matter; the reality underlying the physical has to be grasped intuitively and gradually brought to light.

Can anything be said clearly about this reality? Certainly the rhythmical characteristics of substances and processes, as found in chemistry, are very like the rhythmical periodicities found in music. The repetition of note-qualities at certain intervals, particularly octaves, forms part of the subject-matter of harmony. The periodic system of chemical elements is an expression of the same essential laws.

The correspondence between chemical and musical laws can be traced step by step. Avogadro showed that chemical elements combine in single and multiple proportions. Taking hydrogen as our basis for comparison, we find it has the ratio 1:1 to chlorine in hydrochloric acid. In water, H_2O, there are two parts of hydrogen to one of oxygen. Oxygen is therefore called bivalent. As we have already seen, there are substances capable of forming compounds in various ratios. Manganese is one of these; it forms oxides up to manganese heptoxide. Since oxygen is bivalent, the proportion of manganese in the various oxides, calculated in terms of hydrogen, can be shown as follows:

Manganese oxide	MnO	Proportion of Mn in terms of H	1:2
Manganese sesquioxide	Mn_2O_3	Proportion of Mn in terms of H	1:3
Manganese dioxide	MnO_2	Proportion of Mn in terms of H	1:4
Manganese trioxide	MnO_3	Proportion of Mn in terms of H	1:6
Manganese heptoxide	Mn_2O_7	Proportion of Mn in terms of H	1:7

Other substances, such as iron, favour the ratios 1:2 and 1:3. Phosphorus favours 1:3 and 1:5, sulphur 1:2, 1:4 and 1:6. All these ratios, however, are within one octave, ranging from 1:1 to 1:7. They are found also in the musical intervals of the first, second, third, fourth, fifth, sixth, and seventh. Perhaps it is not inappropriate to express chemical facts in musical terms and to say, for example, that ferrous chloride ($FeCl_2$) sings in seconds, while ferric chloride ($FeCl_3$) is the voice of the third in chlorine iron compounds.

Chemistry may thus properly be called music in matter. Music is an ordering; it brings order everywhere and can order the feeling-life of its hearers, as everyone knows who has felt himself inwardly brought into order by a symphony or some other piece of good music. Music can be shown to have a formative and ordering effect even on physiological processes. More than this: the phenomenon of the Chladnian sound-figures demonstrates the ordering effect of musical sound on matter. This familiar phenomenon is produced by strewing elder-pollen on a metal plate and then playing a note by stroking the edge of the plate with a violin bow. This causes ordered patterns to appear in the pollen dust. These patterns change if the sound is varied. As one thinks of the ordering power of music working right down into matter, one may ask what is responsible for the rhythmic ordering that binds or frees material substances? It is chemical activity. The word chemistry comes from the Egyptian and means 'the hidden'. And chemistry is indeed the music hidden in matter.

Now, looking at a realm even higher than that of music, let us think of going out on a clear, summer night and of the awe we feel as we behold the sublime order of the stars, with the planets making their wonderful curves and loops against the majestic patterns of the constellations. Reverent wonder is the natural response in every human soul. People of earlier times felt this even more strongly, but with a different background.

It is possible today to calculate the paths of the planets for decades and even centuries ahead, to say which planets will be in conjunction or opposition, when and where. The ancients could not do this. But they had an immediate experience of the starry order, which Plato expressed when he spoke of the 'music – or harmony – of the spheres'. In our abstract modern way of thinking we connect quite different concepts with words and phrases, and we imagine the 'music of the spheres' to have been Plato's subjective way of expressing an artistic experience. A more searching inquiry shows that in earlier times people lived in a different state of consciousness, for which the ordered movement of the planets was clairaudiently perceptible as a musical experience. Apart from the fact that the stars imprinted their order on everything earthly and human – the priest-sages of Chaldea, Babylonia and Egypt looked to the stars for the guidance of society and much else – we can easily see that earth-time reflects star rhythms. Minutes, hours, days, months, years, epochs and ages are all measurements taken from cosmic patterns and happenings.

Time plays an essential role in music; thus astronomical laws are again related to the laws of music. Musical intervals and harmonies are an expression of universal rhythms. The relation of the seven planets to the fixed stars, especially the twelve zodiacal constellations, is built on a rhythmical law that finds reflection in the seven intervals and the twelve semi-tones of the scale.

Nowadays it is not customary to speak of seven planetary spheres surrounding the earth. For one thing, the moon is not reckoned one of the planets, but as a satellite of the earth, and the sun is considered a fixed star, not a planet. If, on the other hand, we adopt the geocentric Ptolemaic system and call Moon, Venus, Mercury, Sun, Mars, Jupiter and Saturn planets, then Uranus, Neptune, Pluto and some planetoids are missing from the picture.

If, however, we avoid abstractions and simply go by what we see from the earth, it cannot be denied that from the spatial standpoint Sun and Moon contain the earth within their spheres just as the planets do. And there is an astronomical reason for believing that Uranus, Neptune and Pluto became attached to our planetary system only within relatively recent times.

Looking at the planetary spheres which, from the earth viewpoint, do actually enclose it, we find that an up-to-date understanding of Ptolemy's geocentric system is indeed possible. The otherwise irreconcilable difference between the geocentric and heliocentric systems is overcome from the standpoint of the spheres.

The geocentric and heliocentric systems are probably each correct from a certain angle. Recent ideas about the lemniscate movement of the sun may perhaps justify both concepts. So we feel entitled to speak of the earth and seven encircling planetary spheres.

Stellar movements, music and chemistry thus seem to be varied expressions of one and the same cosmic ordering force. For astronomy to reduce the stellar universe to a mathematically calculable mechanism, as it has done in recent centuries, was just as mistaken as to ascribe colour phenomena to wavelengths or to call the physical body described by anatomy a human being. A higher cosmic order permeates the universe and manifests itself at different levels, down to ultimate physical expression in earth substances. These clearly show the imprint of their starry origin. In this connection we are led to ask – how does cosmic being manifest? We showed above how oxygen, for example, carries the idea of the plant into earthly appearance, and how carbon forms it. In just the same sense that plants are ideas made manifest in earthly matter, every substance is the materialized expression of a process, of a cosmic essence. This essential being dwells in the world of stars, but the world of dense dead matter bears its imprint everywhere.

If we ask why matter has become so dense and fixated, we must cast a look on electricity. We will study a candle flame and note what happens to the burning wax. The hard wax turns into an oily fluid, which is absorbed by the wick and changed into a gas. As the gas burns, heat and light are given off. Here we see a dematerializing process going on, a disappearance of substance. But if the two ends of a wire fastened to an induction coil are placed in the flame, so that an electric current sparks across from one electrode to the other, the process is suddenly reversed. The flame caves in, the production of heat and light stops almost completely, and a black carbon skeleton builds up between the two ends of the wire. This can be taken as a picture of materialization.

Electricity invariably plays the role of a condensing and materializing agent. Anyone who has handled a battery or an induction coil will recall the peculiar odour given off by electric sparks. Some say they smell like phosphorus, or even sulphur. But sulphur has nothing to do with it; the assumption that it has comes from mistaking sulphur-tipped matches for phosphorus matches. But phosphorus also is not responsible for the odour. Phosphorus simply does what electric sparks do: it condenses oxygen to ozone – the real source of the smell.

$$3O_2 \rightarrow 2O_3$$
3 parts oxygen $\rightarrow$ 2 parts ozone

Electricity, then, is a condenser. In contrast to light, which radiates, it is a densifying and materializing force. It can be called light's opposite pole – earth-related counter-light. So we can understand how electricity forces into earthly form the beings and images of which the universe is full.

The process observed in the burning candle shows heat to be an essential phase of both materialization and dematerialization. We described earlier how hydrogen forms a heat-mantle round the earth and how this fire-substance lends wings to the rhythm linking being and appearance. Condensed warmth is the basis of all the phenomena of nature. But what would happen if matter were composed solely of this densified heat? It would be forever vanishing away, incapable of continued existence ·on earth. Electricity endows it with stability.

The concepts offered by atomic physics are extremely interesting in this connection. It postulates atomic nuclei capable of producing heat under certain conditions. But a ring of rotating electrons encircles these nuclei. What is the reality behind such a picture? Heat has been captured and condensed by the flowing electricity; in other words, electricity has bound stellar forces to earth substances.

In this sense we may say that matter is the hieroglyphic writing of the universe. Anyone who trains himself to wide-awake observation of nature's physiognomy can daily experience some aspect of this truth.

CHAPTER TWELVE

Star Patterns and Earthly Substances

The researches referred to throughout this volume, but more especially in Chapter Three, indicate that the organic kingdoms depend for the formation of their very substance on sun and moon, as well as on the relationship of these to the fixed stars. Everyone knows, indeed, that plants, animals and man are all decisively influenced in their physical development by forces from outside the earth. Seasonal changes make this particularly evident. But they are just one of the countless examples that could be given of the way in which the earth is affected by its movement among and its changing relationships with other heavenly bodies. And the whole surrounding universe plays a part in the processes that bring matter into being and again dematerialize it. This is true even with respect to the mineral kingdom. But science has not yet looked into this, so fettered has it been by its adherence to the law of the conservation of matter.

The author conducted a further series of experiments to find out in concrete detail what the effects of these influences were. Cress seeds were analysed to determine how much potassium, phosphorus, calcium, magnesium, sulphur and silicic acid they contained, with the following results:

24 mg. phosphoric acid per gram of seed
18 mg. potassium per gram of seed
7 mg. calcium per gram of seed
6 mg. magnesium per gram of seed
6 mg. sulphuric acid per gram of seed
0·1 mg. silicic acid per gram of seed

This analysis was repeated every two weeks. When the seeds were kept in a tightly corked bottle, the results were constant.

Later, seeds of the same batch were set to germinate in double-distilled water in bowls of rock crystal covered by bell jars to keep the dust out. After fourteen days the plants were 4-5 cm. high. They were then removed from the bowls, dried and burnt, and the ash analysed. Since no minerals of any kind came into contact with the growing seedlings and none were removed, the final figures obtained in the analysis ought to have been the same as at the outset if the law of the conservation of matter was

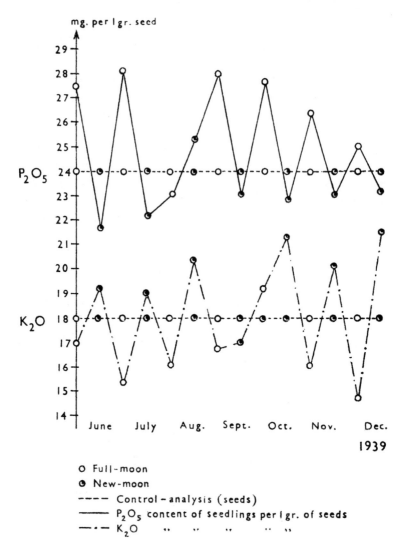

Fig. 5

The emergence and passing away of phosphorus and potassium during the period from June to December, 1939.

68

correct. This was not the case, however. The graphs on p. 68 indicate the variations in the phosphorus and potassium over a period of half a year.

These graphs show that the phosphorus and potassium content of the seeds rose and fell in rhythmic intervals; the emergence of these substances out of non-material states of being, as well as their disappearance from a material stage into an imponderable one, follows the rhythm of the moon phases. But the moon is both an earth satellite and a reflector of the entire cosmos, particularly of the sun and its movement through the Zodiac.

Now, except at one point in each, the two graphs show a characteristic rhythm of even alternation, with the full moon favouring the emergence of substance and the new moon favouring its disappearance. In the phosphorus graph this point falls in August, while in the counter-rhythm of potassium (cf. Fig. 5), it comes in September. It seems that every substance has such a point where the even alternation is disturbed and full moon and new moon exchange their roles. And this point seems to have a relation to the sun's position as it moves month by month through the twelve zodiacal constellations.

This is not to say that the relation of all earth's substances to the constellations has already been experimentally established. Research of this kind has barely been started. We merely point to the possibilities it opens up. Details of further findings made in the course of this experimental work will be communicated later.

We must now try to dispel some of the confusion prevailing on the subject of zodiacal signs and constellations. At the start of our era the sun rose on the first day of spring, March 21, in the constellation Aries. This point on the horizon is called the vernal point. In our discussion of nitrogen we mentioned the fact that it takes the vernal point one Platonic year (25,920 sun-years) to complete one round of the Zodiac. Thus far, in the approximately 2,000 years that have elapsed, the vernal point has progressed far enough to rise in the constellation Pisces. The spring sun no longer rises in Aries, as it did, but in Pisces. Astrology, however, continues to place the vernal point under the sign of Aries, and to say that Aries is to be called the spring sign for all future reference. Looked at in this way, the vernal point becomes a mere convenient device for anchoring a system of co-ordinates. The division of the Zodiac into twelve equal segments, called by the traditional names of the twelve zodiacal constellations, made it necessary to distinguish constellations from the so-called 'signs' thus arbitrarily created. This explains why the 'signs' of the Zodiac are already one whole constellation behind the real positions of the stars and will fall still further behind as time goes on.

We must take into account the fact that the zodiacal constellations vary considerably in size and that the heavens are not divided into twelve equal segments. When in the following pages we associate the twelve months with the twelve constellations, we have only a very rough correspondence in mind.

We showed above how cosmic forces interact in the formation of starch:

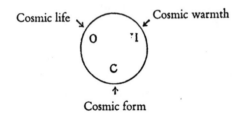

Cosmic life — Cosmic warmth
O ʼI
C
Cosmic form

Starch is the product of cosmic fire, cosmic life and cosmic formative forces. Three substances, hydrogen, oxygen and carbon, are the end-products of its dissolution. We sought traces of their cosmic origin in their physical and chemical behaviour.

Protein may similarly be described as the product of an interweaving of four cosmic principles:

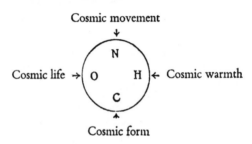

Cosmic movement
N
Cosmic life → O   H ← Cosmic warmth
C
Cosmic form

Now where are these cosmic signposts pointing? What is the origin of these substances which seem to be the basis of all organic nature?

If we follow the activity of hydrogen in the living kingdoms through a whole year's cycle, it obviously reaches a maximum in high summer. The most intensive etherealization of plant substances takes place at this warmest time of the year. Seeds start to ripen, and oil, materialized cosmic fire, begins to form in them.

High summer comes in August, the month when the sun passes through the constellation of the Lion, which has always symbolized flaming courage and all such fiery attitudes of soul.

Old Oriental picturings of lions express this characteristic in a flaring mane which surrounds the head like a golden sun-aura. It may be assumed that the representations of zodiacal figures found in old calendars retain something of the ancient wisdom hidden in pictures. These were the work of priests of olden times, and were never arbitrary, but rather pictorial records of what they knew about the Zodiac. Naturally, these pictures are not to be grasped by a trifling approach, nor are they meant to portray physical animals. They are to be thought of as depicting processes and activities in artistic and pictorial, but at the same time real, images. They were a very real experience to men of old.

The cosmos was recognized by the ancients to have twelve distinctly different phases of activity, and they therefore ascribed twelve characteristics to these heavenly forces. And though we have lost the capacities that conceived these images, we can still feel the truth they symbolize if we approach them with artistic and unprejudiced sympathy.

Leo, the lion, is in this sense the representation of fire-related forces raying from his segment of the universe into all levels of activity and manifestation: the soul-spiritual, the biological, the mineral. Their final manifestation is in hydrogen and its activity; here the sublime process comes to rest. One cannot help thinking of the adage, 'Matter is the last step on the path of God.'

Oxygen, on the other hand, reaches the climax of its activity in nature when the earth is saturated with the fertile moisture of melting snow or mild spring showers, the season when sap begins to stream through roots and stems and there is stirring and germinating everywhere.

This is the moment just before the coming of spring, toward the last part of February, when the sun is passing through Aquarius, 'the water-carrier'. Here is a picture of fertility, the ancient symbol of an activity that blessed and fertilized the earth. Streams of quickening water drench the land and summon into earthly appearance all manner of living forms. This is the opposite pole to Leo's fire-process, and is therefore more to the fore at the opposite season. Oxygen is active at the time when all through nature life is being carried over into physical form with the inflow of fertilizing, saturating water. High summer is the season of oxygen's polar opposite, hydrogen: a subtle fire-process, which everywhere dissolves form and etherealizes matter. These two processes are as opposite in character as the two zodiacal figures Leo and Aquarius are in space.

An objection everyone is sure to raise is that in the tropics, the southern hemisphere and at the poles, plants go through their various life stages at a quite different season. In this connection, let us consider the following.

What might be called the classical cycle of the seasons is a phenomenon of the temperate zones, and so of the more highly civilized parts of the earth. And this normal cycle depends on a balance being kept between terrestrial and cosmic forces. This leads to a harmonizing rhythm between the changing relationships of earth and sun (as cosmic representative). Where earth forces gain the upper hand, as at the poles, or where life is dominated by cosmic radiating forces, as at the equator, abnormal, one-sided conditions result.

Goethe gave the key to a quite new way of looking at all the phenomena of nature with his theory of colour, which cannot be too highly valued. Between the poles of light and darkness live the rhythms inherent in colour; between cosmos and earth arises the wealth of rhythmic life-phenomena of the earth's surface; between poles and equator the rhythmic alternation of the seasons comes into being. The Goethean concept of the two poles and a third new element, rhythm, which reconciles these two extremes, throws fresh light on everything in the organic kingdom. Light and darkness create the rainbow; earth and cosmos create the living kingdom of the earth.

The earth itself is a living organism – a fact still familiar to Kepler. The zone between the poles and the equator is the scene of manifold rhythmic phenomena that mirror cosmic rhythms.

For the earth, the regular succession of the seasons is like human breathing. We have shown how cosmically regulated this breathing rhythm is – a reflection of the sun-rhythm of the Platonic year. But the head and the metabolism of man, which correspond to the earthly polarities of pole and equator, have their own independent rhythms. Therefore cosmic rhythms are mirrored only in the middle region of both man and earth.

Details of such interrelationships are very complex. But if one approaches them with a feeling for the whole picture, many phenomena of earth and cosmos will disclose their secrets.

Now let us return to the question: Where is aerogen (nitrogen) most strongly active? We recall the description given earlier of the forming of protein. It begins with visits of butterflies, bees and beetles to the flowers, where nitrogen's mobility enables seeds to form. This process coincides with the blossoming of plants, the swarming of insects, with the wind carrying pollens of grass and blossoming grain over the countryside.

Such are the main events of the end of May, when the sun is passing through the constellation Taurus. Again, all this is in keeping with the ancient view, which used the Bull as symbol of the forces of motion. Nowhere do we find an old portrayal of a bull lying still and peaceful in a meadow; he is always depicted charging, or otherwise most active. In none of the twelve zodiacal images is movement as much emphasized as it is in Taurus. This sign is clearly the image of universal, all-inclusive forces of movement, which are active at all possible levels, down to the final one: the forming of nitrogen. Thus the zodiacal region from which forces of movement issue is called Taurus, the Bull.

Now the opposite pole of motion is fixity. Just as we showed nitrogen to be the carrier of motion, so carbon was described as the carrier of form. Carbon is nature's great stabilizer. It forms the scaffolding in plants, animals and man, and is the skeleton left after their dissolution. When, in November, the first frosts set the mark of death on the life of nature, when withered remnants of vegetation litter the fields and bare trees look like skeletons in the forest, then comes the time of predominating carbon forces.

This is the time when the sun is travelling through the constellation Scorpio. Forces issuing from this region of the heavens

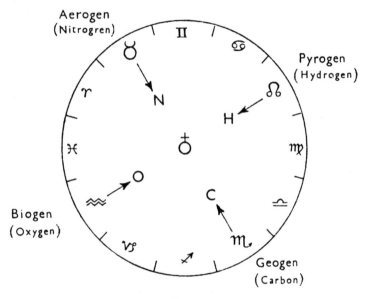

Fig. 6

The atmospheric cross.

have long been symbolized by the scorpion, with its deadly sting and desiccated, skeletal body.

Antiquity had another name for this constellation: the Eagle. This picture was an image of cosmic formative forces operative not in the crudely physical, but on the highest level, that of thinking. It depicted the eagle's power to soar into the sun and survey the scene below from a great height. This symbolized the divine capacity given to man to reflect sublime facts in the thoughts he forms. Materialism was responsible for the eagle being forgotten and replaced by the symbol of death, the scorpion.

Carbon, the last manifestation of the scorpion-eagle forces, also appears in several modifications. The shining diamond is as high above the level of black coal as the soaring eagle is above the crawling scorpion. Indeed, the eagle-qualities noted above have, as it were, been materialized in the diamond's substance. The constellation Eagle-Scorpion possesses a double nature: death in the scorpion, a soaring to loftiest heights in the eagle. We have here a sort of phoenix-motif. But natural death can be looked on in the Goethean sense as a return to essential being, to the disembodied state. When a plant dies and becomes a carbon skeleton, its being withdraws from the material condition into a world we cannot see.

So every death frees a being into higher life. When nature's being remanifests in spring, there is already a seed of death in all its germinating sprouting growth, for without this seed there can be no such thing as shape or form. Involution and evolution, being and manifestation, death and resurrection: all these polarities, which are in turn but a single unity, belong to the nature of the Scorpion-Eagle, the cosmic home of the forces from which carbon issues.

Goethe expressed it:

> Your soul will stay forlorn
> Until you come to know:
> To die and be re-born,
> Is Spirit's way to grow.

A study of these four cosmic principles which breed the material substances hydrogen, oxygen, nitrogen and carbon shows hydrogen and oxygen to be polar opposites in nature, like the constellations Leo and Aquarius. Similarly, nitrogen (motility) and carbon (fixity) belong to the opposite cosmic poles, Taurus and Scorpio. The four together form a cross (cf. Fig. 6).

It will be obvious from the above that the four substances described are reflected in the four Aristotelian elements, fire, water, air and earth. And we should notice the remarkable fact that all four substances are found in our atmosphere, and are indeed, the elements of all the organic kingdoms:

Fire	Pyrogen, or Hydrogen	H	
Air	Aerogen, or Nitrogen	N	
Water	Biogen, or Oxygen	O	Air, Protein,
	Geogen		organic nature
Earth	or	C	
	Formative matter		

The cosmic cross: Leo, Aquarius, Taurus and Scorpio, could thus be called the atmospheric cross, or the cross of the organic cosmos.

These considerations may suggest a possibility of explaining relationships between earth and universe, microcosm and macrocosm, in a way suited to modern consciousness. It is right to repudiate the old, unintelligible conceptions if one cannot arrive at comprehensible, reasonable new ones. New methods of investigation must work at this task. It would be disloyal to the cultural heritage of Central Europe to ignore the meaning and the majesty of the starry order and to fail to sense the deep connection between the heavens above and the earth below. In the work of the great men of our past we find the germs of a new understanding of these interrelationships.

The four constellations chosen for mention in the pages above always enjoyed a privileged position. It is they who give the essence of their being physical expression in hydrogen, oxygen, nitrogen and carbon. Their harmonious interaction produces the substance protein, the basis of all the more highly organized forms of life.

The various properties of protein seem marvellous indeed as one considers how it is formed of these four elements. Each has its own well defined and extensive sphere of action, while together they form a matrix for the development of the three kingdoms, endowed with life (plants), soul (animals), and spirit (man). We came to know oxygen as the element that carries life into physical manifestation, nitrogen as the force that permeates life with feeling and motion, hydrogen as the strongest power of ascent from the material to the spiritual, carbon as the crystallizing agent that brings what is endowed with spirit, soul and life into physical manifestation.

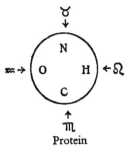

Protein

These four substances in various combinations and permutations enable the creative cosmos to become effective for the earth and to continue reproducing miniature likenesses of the universe in earthly matter. This capacity is most apparent in the seed. Its protein provides the material that brings the prototypal image of the plant into physical manifestation.

CHAPTER THIRTEEN

The Animals

Carbohydrate is the substantiality of plants, protein of animals. The plant proteins in seeds are engendered as a physical precipitation of the commingling in the blossom of plants and animal worlds.

How are these plant proteins distinguishable from animal protein? Chemical analysis cannot always tell one from the other. But many people feel instinctively that there must be a qualitative difference, and find clear evidence of it in digestion. Let us try to answer this question by looking at the nature of animals.

The development of animals from a fertilized cell begins with a completely vegetative process: the cell divides, creating a host of offspring cells (the morula). This takes on a spherical form, the blastula. Then an entirely new phase begins. The blastula starts to 'turn outside in', first becoming indented and then, as the process continues, ending up with an outside sheath which completely closes off an inner space, thus creating an inner and an outer world. This stage is called the gastrula, and it is a decisive one. For up to this point animal development was plantlike. The gastrula stage, however, is the start of a wholly new turn of events. This formation was recognized by Haeckel to be the basic pattern of animal development and the ancestral form of all multicellular animal life.

The 'entoderm' – the interior layer of the embryo – formed by this involution is the primal structure from which the animal's internal organs now develop. The nerves and senses system develop from the outer cell layer, the 'ectoderm'. A wholly new evolutionary principle has come into being with this division into an inside and an outside.

The plant comes into physical manifestation as a direct result of cosmic interaction. In the animal there is an impulse to close off an inner enclave from the surrounding cosmos, and in this enclosure an autonomous system of inner organs develops.

We are familiar with the fact that protein is formed by the addition of nitrogen to the elements constituting carbohydrates.

We need to find out whether this addition does not bring about a complete reorganization of the whole basis of life, such as enables it then to form the gastrula. There is just a hint of such a sheath-formation in plant seeds, where nitrogen is first found in an organic process and produces protein. All seeds form in hollow spaces. This inclines one to see a close relationship between nitrogen and gastrula formation.

We learned to recognize nitrogen as the carrier of mobility. Now let us turn our attention to the plant again. It has its being in light. It is formed by the universe and extends into the starry cosmos. But it is entirely a product of forces outside its organism. Its whole function is simply that of a carrier of life, of life endowed with shape by carbon and then etherealized again by hydrogen. Carbohydrates are thus its characteristic substance.

Something quite different is needed for building animal bodies, which have to house creatures endowed with autonomous movement and sensation. Sensation may properly be called motion of a higher order, soul-movement. Nitrogen acting from outside the organism is incapable of providing the foundation for sensation. A basic interiorizing must occur, so that the animal may become a unified organism capable of free, independent motion. To bring this about, animals take cosmic formative forces into themselves to build up a system of internal organs: heart, kidneys, lungs, liver, etc. The substance plants produce is dependent on external light and subject to direct irradiation by cosmic forces. The protein animals create is an internal product, formed by cosmic forces which have been 'turned outside in'. The internal organs of animals are really an interiorized reflection of forces and processes of the external universe, dynamic foci of an inner cosmos. Such investigators as Paracelsus pointed this out long ago. The diagram overleaf may serve as a chemical-dynamic illustration of the difference between plant and animal protein.

Nitrogen is here again the physical agent that carries out this fundamental movement of interiorization, of turning the cosmos outside in. It is the carrier of motion in that it enables earthly processes to repeat cosmic motions and rhythms in independent organisms. Breathing, as a microcosmic repetition of the sun's

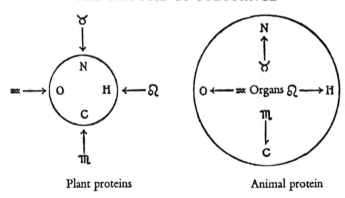

Plant proteins   Animal protein

passage through the Zodiac, was shown to be a prime example of this.

We have touched only on such aspects of these processes as have bearing on the forming of substances. Much else could be learnt about them from other studies.

CHAPTER FOURTEEN

Plant Poisons (Alkaloids)

Carbohydrate is the prime plant substance; plant proteins are found only where animal and plant spheres overlap in seed formation. Blossoms even seem like stationary butterflies, with their fine colour and airy delicacy, while butterflies are like blossoms free to flutter in the air, as Rudolf Steiner said. This image pictures the exchange that takes place between the plant and animal spheres in the most chaste and delicate kind of intercourse, with the protein in the seed as its offspring.

Such is the case with those species of flowers that open fully to the sun. But there are exceptions, species or families which incline more to the animal sphere, and this is reflected in a gastrula-like blossom form. Among these are foxglove, various lilies such as autumn crocus, aconite, henbane, nicotiana and other nightshades. The blossoms of these species are cup-like, in some cases forming hollows almost completely sealed off from the world outside.

The family of Papilionaceae (in German, 'butterfly-flowers') represents a halfway stage between the other two in some respects. Most of them are non-poisonous. But all share one noteworthy characteristic: they form enormous quantities of protein. It almost seems as though their protein-forming tendency were a safety-valve. Perhaps they would be poisonous if they were unable to extrude all this protein.

Now what is it that makes plants poisonous? It is plant protein, the same substance normally produced in seeds when plants are fructified by the animal sphere, but – in the case of poisonous plants – protein that has been denatured through too deep a penetration by the animal nature. When plants go beyond what is plantlike and take over formative processes that properly belong only to the animal realm, there is a corresponding depression of the life-element in the protein formed. Protein is always broken down by a development of conscious feeling and autonomous motion; consciousness is always achieved only at the expense of purely vegetative life. That is why, for instance,

81

nerve substance cannot be regenerated. But protein breakdown is a normal function of the animal body. In the plant, where it is abnormal, it produces poison.

Now what happens when the life-giving elements, oxygen and hydrogen in the form of water, are removed from protein? If no remnant of life remains, the result is cyanide:

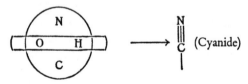

This can be done in the laboratory, using sodium as the protein-destroying element. It is well known that sodium greedily seizes upon the elements of which water is composed; this is why it has to be kept under petroleum. Now a slow stepwise breakdown and suppression of life-elements takes place in poisonous plants, and this process generates the whole list of plant poisons in substances ranging from protein to cyanide. A comparison of the formulae of the various plant poisons with the formula of protein presents a revealing picture of the gradual degeneration protein undergoes. To simplify, we will shorten the protein formula from $C_{720}H_{1134}O_{248}N_{218}$ to $C_7H_{11}O_{2.5}N_2$, and similarly reduce the formulae for the plant poisons, thus:

Protein	C_7	H_{11}	$O_{2.5}$	N_2
Coffein	C_7	H_9	O_2	$N_{3.5}$
Atropin	C_7	H_9	O	$N_{0.5}$
Morphium	C_7	H_8	O	N
Strychnine	C_7	H_7	O	N
Nicotine	C_7	H_{10}	$-$	$N_{1.5}$
Cyanide	C_7	$-$	$-$	N_7

These few examples serve to show the degeneration that takes place between the normal protein stage and cyanide, the deadliest of poisons. This poison-forming process is a gradual dying, especially when animal protein is used as a basis of comparison.

What happens to animal protein when life leaves it? It becomes carrion. The substances this process forms are called ptomaine poisons. They are very closely related to plant poisons.

Greek mythology tells of a magic garden where poisonous plants grow. The autumn crocus still bears a name reminiscent of the site of this garden: 'colchicum'. It was in Colchis, a city on the Black Sea coast below the Caucasus, long associated with the voyage of the Argonauts, that the enchanted garden of Hecate was situated.

The Black Sea is particularly striking for the great contrasts that exist there side by side. Shores that one moment lie sunning under skies of endless blue are suddenly overshadowed by black clouds and whipped by stormy seas. The water contains unusual amounts of both iodine and gold. Harsh cliffs adjoin paradisal gardens. The ancients called this body of water 'The Inhospitable Sea'.

Close by Hecate's sinister enchanted garden, where grew the herbs that brought both death and wisdom, stood a temple of Diana where people went to pray for life, earthly well-being and fertility. They still knew in those days of the deep connection between the development of consciousness and the process of death and breakdown.

Poisonous plants and food plants may grow side by side and even be closely related to each other. Death and life too are polarities, which conceal between them, as a third element, the whole span of human evolution.

CHAPTER FIFTEEN

The Vitamins

During the nineteenth century, nutrition came to be looked on as though it were a problem of physics. This was when the great achievements of physical science were extended into the realm of life through the researches of Liebig, Wöhler and Pettenkofer. As Wöhler himself put it, the synthesizing of urea gave a death-blow to the 'life force' which science had hitherto postulated. The human organism began to be regarded as a heat-powered machine. According to this concept, foods were sources of heat-energy, fuels for stoking the human steam-engine. Carbohydrates, fats and protein were scrutinized for their fuel value, expressed in terms of calories, as though nutritive capacity were simply heat-generating capacity.

Laymen in increasing numbers embraced a popularized form of this conception. Many tried to inform themselves exactly as to how many grammes or calories of carbohydrates, fats and protein they needed daily, and many households felt the shadow of this idea's ghostly presence fall upon them. Shop windows were decorated with the famous drawing of man represented as a factory. The general manager's office is in the head. From this point, all sorts of telephone wires and other transmitting devices carry directives to departments in the lower storeys. An escalator in the oesophagus expedites food, shown as coal, down to the stomach, where the boilers are stoked. Exhaust gases escape from this fabulous building through a chimney-nose.

This conception was put to the test by the Basle physiologist, Bunge, and his associates in experiments with animals. One group was fed on milk, an equal amount to each animal, making sure it had a certain number of calories as fuel. A second group was fed on synthetic milk, a mixture of milk proteins, milk sugar, milk fats, and salts. Both groups would have got along equally well if nutritive and caloric values had really been as identical as the prevailing view assumed. But the opposite turned out to be the case: the animals fed on fresh milk grew up lively and happy, while those fed on synthetic milk died off.

84

THE VITAMINS

The logical intellect of the time (1882) now came to the conclusion that there must be some ingredient in fresh milk that had escaped detection by chemical analysis. This ingredient must be a very complex substance, but with more perfect and refined methods of analysis it would eventually be isolated. Because of its importance to life, this hypothetical substance was later (1912) called 'vitamin'.

A tidal wave of research now got under way, followed by a flood of literature on vitamins too extensive to keep track of. Diseases (avitaminoses) caused by vitamin deficiencies in the diet began to be recognized and tagged with A, B, C and D. This led to assuming the existence of vitamins, which were then given the same A, B, C, D tags. Later on, E, F, G, H and K vitamins were added to the list. More recently, however, a renewed conviction has developed that these latter vitamins are somehow related to the first-named group; that will be clear at the end of this chapter.

Despite all the zeal and intelligence spent on vitamin research, vitamins remained more or less a problem. All sorts of substances were analysed as possible vitamin carriers. A typical case was that of Vitamin C. Famed vitamin research men were finally able, after years of painstaking work, to produce by fractionation an extract that had an enhanced vitamin action in proportion to its weight. But when they tried to isolate a definite substance in this plant-juice concentrate they arrived at quite differently structured substance. One scientist – the Norwegian O. Rygh – described his product as a narcotine derivative, while the other – the Hungarian Szent-Györgyi's – was an aliphatic substance related to sugar. But both were claimed to have a connection with Vitamin C.

The situation was thus one in which two substances, each said to be a direct or indirect carrier of Vitamin C, had nothing in common chemically, as the formulae on p. 86 show.

The same sort of story, with two scientists searching for the same vitamin and coming up simultaneously or in succession with completely different substances, can be found repeatedly in vitamin literature. It seems as though vitamins elude the chemical approach. Studying the literature, one gets the impression that

85

vitamins are entities at a level above matter, forces active in matter rather than themselves material. The so-called synthetic vitamins now on the market are perhaps optimal carriers, but they are full of surprises when it comes to therapeutic effectiveness or

Vitamin C

according to Szent-Györgyi
(1932)

according to O. Rygh
(1932)

Ascorbic acid

Methyl nor-narcotine

non-effectiveness. This shows that the question of vitamin carriers is by no means fully solved.

The following picture might serve to illustrate the situation. A gramophone record can be made of hard rubber, of synthetic resin, or of a number of other materials. Its composition is not the important thing. What is important is the music, which is graven into the record, and can be reproduced under suitable conditions. Similarly, the chemical formula of a vitamin-bearing substance is not as essential as the force that seeks out or builds itself appropriate substances to work in.

Now what is the nature of the forces in Vitamins A, B, C and D?

It may be helpful here to draw a Goethean picture of what he might have called the physiognomy of the avitaminoses. Only the outstanding features need be characterized for this purpose. If they are correctly drawn, smaller details can be fitted into the whole picture easily enough. We must take note too of how the natural substances that carry vitamins fit into nature's living processes. Knowledge of these interconnections will not be furthered by studying isolated vitamins, but rather by a survey of the total complex of Nature's elementary processes.

It is said that Vitamin A is contained in oils, fats (especially fresh butter), fruits and blossoms. There is an essential connection between these various substances. Fruits and blossoms are products of warmth processes predominant in plants at the warmest time of year. Oils are the clearest expression of warmth qualities, and in an earlier chapter we therefore called them materialized cosmic fire. From the chemical standpoint they are simply carbon skeletons clothed in a garment of hydrogen. Plant oils burn and thus reveal their latent fire. Animal fats and oil have the same character in their own special way. Whales, seals and other Arctic mammals are enclosed in an armour of fat that regulates the warmth organism of these dwellers in the icy depths.

What may we say is warmth's essential nature? We see it in its most primal form in the warmth-element of plants, reaching out to the highest levels of the summer atmosphere. We described how the plant-being floats up on wings of hydrogen to the outermost boundaries of the earth's air-mantle. All other physical warmth-phenomena repeat this process. Gases, fluids, metals – all heated substances expand. Expansion is as characteristic of heated minerals as growth is of living organisms under the influence of warmth.

It is well known that the chief symptom of the illness resulting from a deficiency of A vitamins in the diet is stunted growth. Further progress of the disease results in atrophy, especially of the eyes, and in other disturbances on the periphery of the organism, the epithelium. Stunted growth is, however, the most conspicuous effect.

When diets lack the stimulus supplied by oils, fats, fruits and blossoms, the organism fails to generate warmth and growth is inhibited. Quite clearly the periphery is the first part to suffer from such deprivation; the skin and sense organs begin to show unhealthy changes. We venture the opinion that Vitamin A is not just this or that chemical substance, even though it is dependent on matter for support, but, rather, living warmth that has created an organic precipitate in fruits and blossoms, oils and fats.

Vitamin C, on the other hand, is found in leafy vegetables. Indeed the green leaf is the basis of a plant's existence, for through it light is assimilated. Leaves and their nascent starches are creations of light; they are organized and formed by light. In an earlier chapter we described plant substance as an enchanted rainbow, a rainbow that reappears in the blossom colours as the plant reaches a peak of upward growth and is on the verge of dissolution. We can follow this metamorphosis of light in plant growth from its first material appearance in the leaf to its radiant release in starry blossoms. The green leaf may be called latent light.

The avitaminosis due to lack of green vegetables in the diet is scurvy. The immediate impression an open mind gets of a person afflicted with this disease is that he has a light-starved organism. One can see what this means by looking at a plant grown in the dark. It is 'spineless', floppy, lacks structure and wilts quickly. There is no body to it, only a long pale shoot. The same deprivation in animals and man naturally produces somewhat different symptoms. We may recall the gastrula formation which, in the higher kingdoms, has interiorized previously external forces, so that the light from inside out is met by a light-process within.

Man's skin is the organ which holds a balance between internal and external light. Healthy persons maintain this balance; they show it in a rosy complexion with a bloom on it and in a good skin texture. The light that shines from the eyes of persons dear to us is also inner light.

Now when outer light gets the upper hand, the skin turns brown to protect the organism. But the balance can also be similarly disturbed by a lessening of inner light, in which case one gets an impression very like that of a weak, wilted plant. Persons ill with scurvy also have a yellowish or brown skin colouring, and the balance can be disturbed to a point where the skin gradually breaks down. Poor texture permits bleeding, and organs related to light-processes in the organism, such as the kidneys and adrenal glands, are also affected.

Thus, neither narcotine nor ascorbic acid is the essential element in Vitamin C, but the living light materialized in green plant leaves. And when this light latent in green portions of vegetables is lacking in a diet, there is nothing to stimulate a radiating out of inner light in man's organism.

Vitamin B is absorbed from eating the skins and hulls of fruits and grains, especially unpolished rice.

It is an interesting part of a study of civilization to observe for how long nations and even whole continents have to suffer for lack of human insight. Rice is the staple of East Asian diets; a large percentage of the population eats almost nothing else. There were no nutritional problems when rice was eaten whole, just as it was harvested. But as European civilization became increasingly influential and rice was put through modern mills and de-hulling devices, the Malayans were one of the peoples unable to get anything but polished rice. Beri-beri was the result, though at first its nature was not understood. Untold numbers were felled by the disease. At first it was thought to be an epidemic, some sort of plague. It chanced that a Dutch physician in charge of beri-beri patients in the Dutch East Indies owned a chicken-farm. The chickens were fed on polished rice. One day the doctor had to recognize that they were falling ill of a disease with symptoms similar to those of beri-beri. When this had gone on for some time, bran was fed to the chickens one day when no rice was available. Surprisingly, the chickens recovered. Rice-bran extract was administered with similar success in the hospitals, with the result that beri-beri came to be recognized as a deficiency disease.

No catastrophe ensues when Europeans eat polished rice, for they often eat black or brown bread, and these contain parts of the grain hulls. Apples are often eaten with the peel, as are other fruits. It is the Asian, subsisting almost exclusively on rice, who becomes ill when it is polished.

It is obvious, then, that forces of some kind are hidden in the sheath. In order to find our way into this problem, let us picture boundless reaches of undifferentiated space and then imagine rounding this space off into a great sphere with one circling, shaping gesture. Would this not be the first step in the creation of an ordered cosmos?

Just such a gesture is depicted in all myths dealing with the creation of the universe, when they describe the first act of Divinity as the creation of heaven and earth. The creation of the macrocosmic vault of heaven is the same act in infinity as the creation of an enclosing boundary or shell round the finite microcosm of the earth. It is impossible to conceive 'space' without boundaries; no matter how far out we extend the limits, the concepts space and spatial order still contain an element of finiteness. Indeed, it is a big question whether cosmic space can be grasped at all with our physical concepts of the spatial. Perhaps the ancients, who pictured outer space as an enclosing shell with the fixed stars attached to it, really had hold of an important truth: the fact that an ordering of space can be achieved only within a boundary. It is perfectly possible to conceive universal space as a sphere, the boundaries of which reflect cosmic activity, without thinking of these boundaries as physical. This is the scene of the ordered star movements experienced as 'the harmony of the spheres' by ancient man. We described how these patterns work as music right down into the chemistry of material substances; how star patterns, music and chemical action are expressions of one and the same universal force whose archetype is the 'shell', the enclosing sheath of the cosmos. Sheath forces are ordering forces, or chemical action.

Now if there is a lack of these forces in the diet, the inborn chemistry, the primal inner order, of the organism, lacks stimulation, as becomes apparent in the first symptom of beri-beri. The muscles of the ankle lose the power to contract. Examination shows that they have lost their normal structure or patterning. The muscle fibre gradually dissolves and becomes a pulp; paralysis sets in, and the nerves degenerate. That is why Vitamin B is called an anti-neuritic.

To repeat: Vitamin B is not this or that chemical substance, even though it makes use of some such material, but rather an ordering force at home in sheaths; in chemistry, inner structuring.

Vitamin D, lastly, is said to be present in fish-liver oils, and phosphorus and sea salt both carry it. It would be hard to understand why fish-liver oils figure in the picture if we did not know that cholesterol and lipoids formed by the liver are present there in solution. We know that these substances join forces with carbon to build the framework of the whole organism; they are the material of which supporting tissue and cell membranes are made. It is quite correct to picture them as stages on the way to bone-building. In arteriosclerosis, the walls of the arteries become coated with cholesterol, which gradually calcifies.

Now cholesterol and lipoids have phosphorus in them, and phosphorus gains added interest from this fact. Later on we will describe in detail how this substance functions as a densifying, mineralizing force in nature.

Salt that has just crystallized out of solution gives us an archetypal picture of how the skeleton is formed. The human bony system 'crystallizes' in the fluid embryo in the same way that salt cubes take shape in the mother-solution. The details of this process are naturally very complicated, but in the end it produces the hard, solid mineral substance of the bones. The forces at work here are densifying, shaping ones.

When these are lacking in the diet, the organism's primal formative forces lack the necessary stimulation, and the skeleton is not properly completed. The resulting deficiency disease is rickets.

Vitamin D, also, is not simply a chemical, but a shaping universal force whose archetype is crystallizing salt.

	Essence	Carrier	Avitaminosis
Vitamin A	Warmth	Oils	Stunting
Vitamin B	Order	Hulls anu peel	Beri-beri
Vitamin C	Light	Leafy vegetables	Scurvy
Vitamin D	Form	Lipoids	Rickets

This table will bring to mind an earlier one in which the four substances hydrogen, oxygen, nitrogen and carbon were shown as reflections of the Aristotelian elements. We can perhaps suggest that the four vitamins are related to the same four substances at a higher level. They may be conceived of as energies not yet in a state of material fixation, forces capable of forming complete healthy protein by their harmonious interaction.

Some time ago, the author did some experimenting to verify what has been said above about the nature of vitamins. The experimental set-up had to be adapted to the nature of vitamins and their fluctuating radiation.

When light rays are passed through a prism, the spectrum of colours appears on a white background. This visible spectrum passes over into an invisible extension at each end. At the red end is infra-red, whose presence can be determined with a thermometer as warmth. At the violet end is ultra-violet, which, though invisible, registers its presence chemically on a photographic

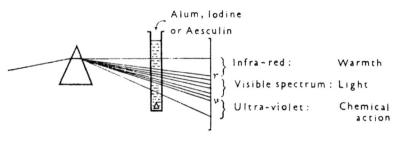

Fig. 7

The three parts of the spectrum: Warmth, Light, Chemical action. According to the content of the absorption, Vessel A, one of the three parts is extinguished:

Through alum solution – Warmth (infra-red)
Through iodine solution – Light (the visible spectrum)
Through aesculin solution – Chemical action (ultra-violet)

plate. But we do not see the thing photographed. The 'picture' that emerges after the plate is developed is due to a chemical reaction caused by the invisible ultra-violet rays.

If a vessel containing a solution of alum is interposed between the rays and screen, the warmth end of the spectrum disappears. Infra-red is as though swallowed up, while the light and chemical rays pass through unhindered. But if a vessel filled with an iodine solution is interposed, all the visible light rays are absorbed, while the chemical and warmth rays pass through undisturbed. In a previous chapter we described iodine as a light-thief. If, finally, we interpose a vessel filled with a solution of aesculin, the glucoside contained in chestnut hulls, the ultra-violet chemical rays of the spectrum disappear, though the light and warmth rays go through as before.

This method makes it possible to distinguish between the

cosmic forces combined in the spectrum and provides an experimental set-up in which any one of the single components can be excluded. The interposed vessel is replaced by a double-walled glass sphere, with a space between the inner and outer sphere to hold the test solution. Three experimental fields are created by filling this space in turn with solutions of alum, iodine and aesculin, to shut out the rays of cosmic warmth, light, and chemical-formative action.

We still lacked an experimental means of excluding the fourth cosmic energy, the formative forces. These are not present in the earthly spectrum. But Goethe's colour-circle shows a heavenly counterpart to the earthly one in the space between red and violet. In the earthly spectrum, green is found between blue and yellow. Goethe's colour-circle shows heavenly purple, or peach-blossom hue, opposite green. His idea of arranging the colours in a circle, or in two triangles, at once brought life and meaningful order into the whole; the linear spectrum conceals the real nature of colour, which hovers between the material and non-material.

This purple cannot be made by mixing red and violet as one mixes blue and yellow to make green. Goethe produced peach-blossom by letting the red end of one spectrum cross the violet end of a second one. There is not the mixing of substances involved in making green of blue and yellow, but the much subtler action of interpenetrating insubstantial rays. Purple is thus in a special category. When it is lightened, it becomes the rosy colour of the human skin, 'incarnate'. This hue is undoubtedly the highest intensification and synthesis of which colour is capable, a colour which is not fixed but exists in living interplay. It constantly changes with the changing relationship of soul and body. When the soul leaves the body, this rosy hue at once changes into its opposite, a greenish hue. Peach-blossom, or 'incarnate', is thus a relative of purple.

We know that the linear spectrum has no boundaries, that the visible line extends into infinity by way of infra-red in one direction and of ultra-violet in the other. But mathematics tells us that at the cosmic periphery the two infinities are the same. This might be pictured for earthly eyes in the following diagram:

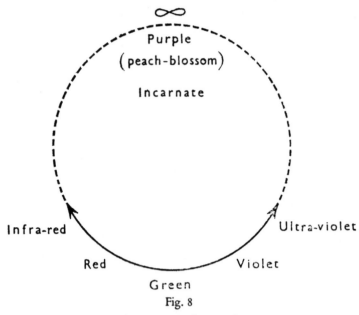

Fig. 8

The Cosmic Colour-circle

The visible spectrum – laid out on the flat – would if extended to left and right beyond the infra-red and the ultra-violet, stretch away to infinity. But the infinity to the left is the same as the infinity to the right; in the closed circle of the diagram it lies opposite earthly green as the place of the 'incarnate'.

Here, as we see, purple lies across the circle from the green.

Now the question is how to bring this celestial purple within earthly testing range. In this cosmic sphere everything that holds true under earth conditions loses validity; things are reversed. On earth, space is filled with something; its reverse is the vacuum. Perhaps a vacuum between the walls of our double sphere, where we had put solutions of alum, iodine and aesculin, would create the right kind of experimental field for the exclusion of formative radiation.

We should emphasize that the experiments described here were concerned with living forces that are not active only in organisms. We wanted to determine their independent existence as cosmic forces.

The following consideration throws more light on the relation of the vacuum to the formative forces. A vacuum is an empty

Fig. 9
Normal crystallization.

Fig. 10

Crystallization in a Vacuum.

Square and crescent as 'rickets images', the crystalline rays are formed directly
the vacuum is released.

Fig. 11

Crystallization in a vacuum countered by phosphorus (or cod-liver-oil). 'Rickets cured'.

Fig. 12

'Rachitic picture' in a vacuum which the phosphorus can no longer overcome.
(Crystallization after 12 hours.) The salt in the solution has been brought to the
stage of forming a square and crescent; because of this, no crystalline rays
emerge when the vacuum is released (see fig. 10).

space with the power of suction. But the force that exerts the suction is the same contractive force, usually called cohesion, which we must imagine throughout space. What is it that keeps the terrestrial globe from disintegrating into the surrounding universe if not the force of suction which we ordinarily call gravity, but which is as much an expression of cosmic shaping forces as cohesion is?

In experiments with formative phenomena in the vacuum sphere, one would expect some sort of interference, if the above assumption is correct. For just as warmth is devoured by alum, chemical action by aesculin, and light by the light-thief iodine, so one might expect formative forces to be swallowed up by a vacuum. And it was indeed possible to establish the fact that crystallizing salt, the archetype of formative process, behaves quite abnormally in the vacuum sphere.

In crystallization studies the usual procedure is to examine how edges, angles and surfaces are formed in single crystals. But the relative positions of single forms crystallizing out of a solution are also an important aspect. When, for example, a saturated solution of potassium nitrate is allowed to stand in a crystallization dish, the needles crystallize in about ten minutes in the form shown in fig. 9.

The composition here is somewhat chaotic, with no clearly marked orientation.

But when the crystallizing is done in an experimental field surrounded by a vacuum, no crystallization occurs for quite a while. Only some hours later do we see small round amorphous deposits on the bottom of the dish. As time goes by, these little piles of soft powder form themselves gradually into distinct squares and crescents. Directly the vacuum is removed and air pours into the space between the walls of the sphere, crystalline rays shoot out from these formations (cf. fig. 10).

These formless little piles, apparently the product of a lack of formative forces, were dubbed 'ricket crystals'. This term simply expresses the fact that an exclusion of cosmic formative forces calls forth abnormalities of form which, when the same thing occurs in human organisms, cause rickets.

If this assumption is correct, however, the addition of substances known to contain Vitamin D, such as phosphorus and cod-liver oil, might be supposed to reverse this abnormal tendency. Experiments along these lines were unsuccessful so long as cod-liver oil or phosphorus was added directly to the salt solution. It was only when resort was taken to the radiating action of these substances that a 'healing' of the rickets crystals was effected.

In this experiment, a three-walled sphere was substituted for the double-walled one previously used. The vacuum sphere was on the outside, as before, while a highly dilute phosphorus solution filled the adjacent inner space.

In this set-up, the potassium nitrate solution took less than ten minutes to make the fully formed and ordered pictures of 'healed' rickets shown in fig. 11.

When the experiment was repeated with the same phosphorus solution, more and more time was required for crystallization to take place, and the crystallization picture grew progressively more chaotic. This meant that the phosphorus was gradually being exhausted. When it came to the seventh experiment and after a wait of several hours, there was still no sign of crystallizing. Everything (including the vacuum) was left standing overnight. The next morning there were again little squares and crescents of powdery sediment (cf. fig. 12).

The vacuum sphere, then, has the power to prevent the ordering, formative forces of the cosmos from structuring true crystals. But one has to picture a certain minimal irradiation of the salt solution. It is too weak to bring about a true crystallizing, but is just able to form squares and crescents in the soft powdery mass of sediment. These forms struck me as rather like visiting cards, so to speak, which the cosmic formative forces were delivering through the slit in a locked door.

The square is the simplest possible expression of purely mineral formative forces. The three-dimensional cross in space is represented in its purest form by the cube. Where in the plant kingdom one finds this formative force most strongly active, a cross-section of the stem reveals a square patterning. The labiates, which do not breathe out their fragrance, but conserve it in an aromatic stem and foliage, are an example. In the case of plants such as marsh marigold and coltsfoot, which incline especially to the watery element in their make-up, the edges of the leaves have crescent-shaped indentations. The half-moon form is the basic shape created by the ordering force of chemical activity present in the fluid element.

The exclusion of cosmic formative forces thus called forth an abnormal process comparable to that which produces rickets in the human organism. This process was reversed when substances known to contain Vitamin D were added and took effect. This completes the proof that Vitamin D is not a chemical compound that can be synthetically produced, but a primary cosmic formative force.

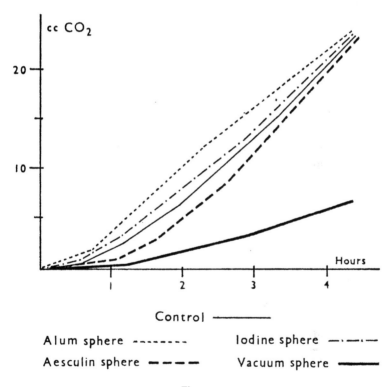

Fig. 13

Yeast fermentation (producing carbonic acid) in spheres of alum, iodine, aesculin and vacuum – i.e. with Warmth, Light, Chemical action and formative forces excluded in turn.

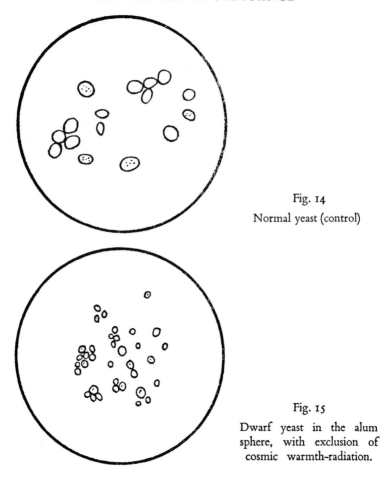

Fig. 14
Normal yeast (control)

Fig. 15
Dwarf yeast in the alum sphere, with exclusion of cosmic warmth-radiation.

Suitable subjects were similarly studied inside spheres of alum, iodine and aesculin. Yeast proved to be very good for this purpose. It is an organism with an easily observable life-process. It breaks sugars down into alcohol and carbonic acid. The carbonic acid can be measured, and thus yields information about the life-processes in the yeast. Fig. 13 shows the rate of carbonic acid formation in yeast subjected to the radiation of the various spheres for equal time periods.

Yeast grown inside the alum and iodine spheres, and so without cosmic warmth and light respectively, produces a greater car-

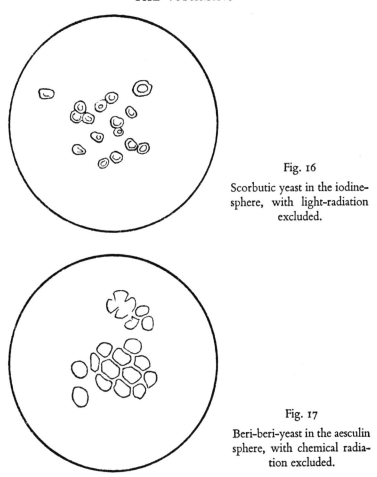

Fig. 16

Scorbutic yeast in the iodine-sphere, with light-radiation excluded.

Fig. 17

Beri-beri-yeast in the aesculin sphere, with chemical radiation excluded.

bonic acid development compared with a control in which only water was used to fill the sphere. Yeast grown inside aesculin and vacuum spheres, and so without cosmic chemical action and formative forces respectively, shows a greatly reduced production of carbonic acid. But this large deviation from the control curve can be annulled by adding substances known to contain Vitamins A, B and C.

In the first case, an emulsion of butter was used; a drop of it was added to the ferment. An extract of spinach was mixed with the abnormally developing yeast inside the iodine sphere. A drop of rice-bran extract in the aesculin sphere reversed the abnormal development there. The abnormal carbonic acid curve resulting from the use of the vacuum sphere could be normalized only by radiating the energy of Vitamin D carriers through the vacuum into the yeast.

Microscopic pictures of the yeast show even more clearly than the graph the kind of abnormality that results from excluding one or other of the cosmic energies. Fig. 14 gives a microscopic view of normally developed yeast.

Fig. 15 shows the dwarfing that came about in the alum sphere, where cosmic warmth was excluded. But a glance at the graph (cf. fig. 13) shows that an abnormally active production of

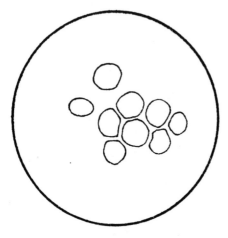

Fig. 18

Rickets yeast in the vacuum-sphere, with cosmic forma-tive forces excluded.

carbonic acid went on there. This gives an accurate picturing of the deficiency disease caused by a lack of Vitamin A.

When cosmic light was excluded by the iodine sphere, the organisms that developed looked squeezed out and hollow at the centre: dry, wilted forms which exuded a slimy substance into the developing fluid. When this fluid was analysed, it was also found to contain an abnormal amount of acid. All this adds up to a picture of scurvy. (Fig. 16).

Fig. 17 shows yeast grown in the aesculin sphere, which ex-cluded cosmic chemical action. The cells here are relatively large, but have weak contours and lack interior structure. Where they touch, their edges flatten out, giving the impression of a conglomerate much like honeycomb. Through the microscope one sees cells bursting apart or dissolving into the surrounding liquid. As the graph shows, carbonic acid development is strongly inhibited. The whole picture is one of autolytic disintegration, just the condition found in beri-beri.

In the vacuum sphere, where cosmic formative forces are excluded, we find huge distended, globular organisms (cf. fig. 18). There was no trace of nuclei, and here again we observe a soft, amorphous honeycomb-conglomerate. Carbonic acid development is extremely sluggish. In some experiments it came almost to a dead stop (cf. fig. 13). This is a picture typical of rickets.

These morbid forms, produced by excluding the four cosmic nature-forces, were rectified in every case by adding or otherwise activating the corresponding vitamin-carrying substance.

This completes the characterization of these four universal forces and supports the view advanced here about the nature of vitamins. It was mentioned above, that nowadays many more vitamins have been found in addition to the classical A, B, C, D. Indeed, in future, there must be found as many vitamins to exist as living organisms in plant and animal. Hence, these four universal forces are active and interweaving in every living organism, revealing more of one or the other side of their activity. It will not be easy to distinguish their nature all the time, but approximately the relationships to the great Four will be found.

The experiments that gave these unmistakably clear results were carried out by the author in the summer of 1929, in collaboration with G. Suchantke, of Berlin. Later experiments, carried out elsewhere, in 1931 and 1932, gave less striking results, but the first results were statistically confirmed. Experiments of this kind are influenced by even slightly changed circumstances which do not ordinarily have to be considered, such as whether they are done in a basement or on a higher floor. A further extensive series of experiments is needed, to throw fresh light on all the problems that arise in this work.

CHAPTER SIXTEEN

Coal Tar Chemistry – Realm of Mirror-images

W e described in earlier chapters how richly varied a spectrum of substances develops out of the virgin substance, starch, during the plant's growth and flowering, while rootward contraction hardens the starch into structural cellulose. This woody substance is the lasting element in plants. Roots, together with everything in the root zone, have largely emancipated themselves from the rhythm of the seasons. Leaves and blossoms wither and decay, but the root goes on living through the winter and enables new life to sprout forth in spring. But even when the whole plant dies, its cellulose components maintain a mummified existence. Certain meteorological and geological conditions favour this mummification process, as we see in the resultant layers and deposits of coal and lignite, formed in long past ages of the earth.

Coal, which formed in very ancient times, is older than lignite. The vegetation that became coal cannot, of course, be compared in appearance to plants of the present. And we will leave aside the further question as to whether ancient vegetation was of as dense and hard a substance as that found in the plant kingdom of today.

Now when coal is coked – i.e. heated in an air-tight oven instead of being burnt – it smoulders, and a certain amount of gas escapes, the amount depending on the type of coal. Illuminating gas was produced by this process, first in England over a century ago, and then everywhere on the Continent. This gas is a compound of hydrogen and various hydrocarbons: methane, ethane, ethylene, acetylene, and other such.

Coke is the by-product of gas manufacture. It is used not only as domestic fuel but for smelting iron ore; it is essential for the making of steel.

A by-product of the coke and gas industry is coal tar. At first it was just a nuisance, until ways of making use of this or that component were hit upon. Now coal tar has become the mainstay of the modern chemical industry.

Fractional distillation of tar produces not only certain solids, but also distillates ranging from components with a low boiling point (light oils) to medium-heavy and heavy oils. Their chemical composition varies with the type of coal used.

Paraffin is the chief product of lignite tar, whereas coal tar is composed almost wholly of cyclo-paraffins. This is a chemical distinction essential to a full understanding of what follows, so it requires some explanation. Both groups are chemically inactive. This quality accounts for their title, for paraffin comes from 'parum affinis' – without affinity. And these substances are truly so dead as to resist even the strongest reagents. They undergo no change at all, even when boiled with an acid as strong as concentrated sulphuric acid.

Chemical analysis shows their components to be carbon and hydrogen – i.e. hydrocarbons. The structure of these hydrocarbons varies, however, according to whether they are derived from lignite or from coal. We pointed out above how carbon's prodigious capacity to create form appears in the fact that carbon with its valency of four, can combine with itself. This makes a huge number of combinations possible. The paraffins are hydrocarbons with a so-called open chain structure, while the cyclo-paraffins are cyclic hydrocarbons, with a hexagonal ring structure.

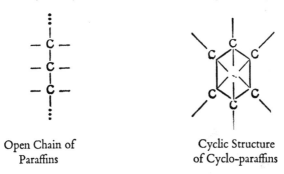

Open Chain of
Paraffins

Cyclic Structure
of Cyclo-paraffins

These structural differences between paraffins and cyclo-paraffins determine their further value for chemical purposes, as will be seen below. Meanwhile, we append a diagram which does not pretend to be complete but gives an approximation of the various distillation products.

Mineral oils (raw petroleum) have a composition similar to

Boiling-point	Lignite-tar	Coal-tar	
	Paraffins	Cyclo-paraffins	
60° to 85°	Benzine $CH_3 \quad CH_3$ $(CH_2)_4 \quad (CH_2)_5$ $CH_3 \quad CH_3$	Benzol	Instead of the complete formula the abbreviated formula for benzol will be used
100° to 200°	Petroleum $CH_3 \quad CH_3$ $(CH_2)_7$ to $(CH_2)_{15}$ $CH_3 \quad CH_3$	Toluol	Xylol Pheno
200° to 300°		Naphthalene	
300° to 350°	Paraffin oil Lubricating oil Vaseline $CH_3 \quad CH_3$ $(CH_2)_{16}$ to $(CH_2)_{18}$ $CH_3 \quad CH_3$	Anthracene	Phenanthrene
above 350°	Paraffin $CH_3 \quad CH_3$ $(CH_2)_{20}$ to $(CH_2)_{24}$ $CH_3 \quad CH_3$	Fluoranthene, Chrysen, Retene etc.	

that of lignite tar. Here too, fractional distillation yields products of the paraffin group: benzine, petroleum, paraffin oil and paraffin. Raw petroleum could be called a natural tar that may have been formed when some great catastrophe of our whole planet kept carbonized plant remains smouldering on.

Lignite tar began to be used relatively early in motor fuels, lubricants and paraffin, thus becoming a chief contributor to the rapid advance of industrial technology. Until well into the middle of the nineteenth century, coal tar was a most unwelcome by-product of gas and coke factories, laying waste the surrounding countryside. It was 1846 before the bactericidal, life-destroying properties of coal tar were recognized and it was put to use as a preservative, first of railway sleepers, and then of wood for every use.

The reason for the neglect of coal tar was its resistance to chemical and oxidizing agents. Coal tar derivatives do not burn as easily as lignite tar products, and thus cannot be used to fuel internal combustion engines. Moreover they are not viscous like paraffin, and hence cannot be made into lubricants.

This deadlock was broken with the discovery that a combination of sulphuric and nitric acids could break down the cyclo-paraffins. Treatment of benzine with this mixture yields nitro-benzine, sometimes called false oil of bitter almonds. This very aromatic substance is used to scent floor wax, shoe polish and the like. This may be why the wide field of cyclo-paraffin chemistry is called 'aromatic chemistry'. Open-chain chemistry, on the other hand, is called 'aliphatic chemistry' because careful oxidation can convert these products into substances with properties resembling those of the fatty acids (aliphatic= fatty).

The nitrification of benzine was a revolutionary advance. The chemical industry grew to giant size. Today the number of synthetic derivatives must be reckoned in millions.

Aniline is produced by treating nitro-benzine with hydrogen:

$$N O_2 \longrightarrow N H_2$$

Nitrobenzol   Aniline

The use of very complicated processes to achieve the fixation of nitrogen led to the first synthetic coal tar dyes:

$$N H_2 \xrightarrow{+ HNO_2} N = NOH \xrightarrow{+ Aniline} N = N \qquad N H_2$$

Aniline   Diazobenzol   Aminoazobenzol
                                   (aniline-yellow)

Perkin put the first aniline dye on the market in 1856. Baeyer succeeded in making a synthetic indigo that could be factory-produced. This was soon followed by the indanthren dyes and the whole range of other dyes now commercially available.

It took tremendous effort and incredibly acute minds to achieve this impressive technical progress. Great self-sacrifice and strokes of destiny marked the paths of the scientists responsible for it. It is very moving to follow the life-story of Runge, for example. Despite great personal difficulties he kept working on the coal tar problem, until – in 1834 – he succeeded in getting a blue reaction-product by treating certain coal tar distillates with calcium chlorite. But it took two more decades for the ingenuity of Hofmann and others to discover how to make technical use of Runge's findings.

Between 1858 and 1865, Kekulé founded modern structural chemistry, thus making a significant contribution to research in the field of the new dyes. The following incident illustrates the part destiny played here: Kekulé fell asleep while riding on the upper deck of a London bus and missed his stop. This was compensated for by a dream in which carbon atoms joined hands and danced around him in a circle. This dream inspired the creation of his structural chemistry.

It was soon discovered that substances of this group, produced at intermediate stages of dye manufacture, react on the human organism. This ushered in the era of synthetic drugs. Salicylic preparations such as aspirin were produced from phenol.

| Phenol | Salicylic acid | Aceto-salicylic acid (aspirin) |

This was the start of chemical therapy, which sought to exploit the discoveries of the dye chemists. Just as dye colours can be changed and even determined in advance by adding side chains to the benzine nucleus, chemical therapy starts with a nucleus able to react on the living plasma and varies the effect by adding side chains. This was how Salvarsan originated. Its scientific name and formula are as follows:

Para - dioxy - meta - diamino - arseno - benzol - hydrochloride

There were sweet-tasting substances as well among these intermediate products of dye manufacture. The sweetest were at once synthesized and put into systematic production. Such was the origin of saccharine, dulcin, and other such sweeteners.

| Dulcin | Saccharine |

Nitro-benzol, the primary substance of this whole development, is aromatic in character. Here too, ingenious synthesizing yielded an abundance of other artificial scents. This led to the development of the whole synthetic perfume industry with its immense variety of compounds, each imitating a different plant fragrance.

No one can dispute that this expansion of coal tar chemistry is a triumph of the human mind. But what spirituality was

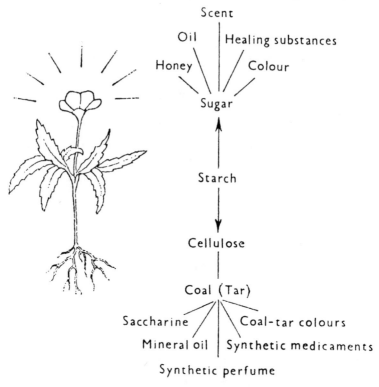

Fig. 19

The sub-earthly spectrum of substances as a mirror-image of natural substances.

behind it, and how are these substances related to the cosmic whole?

Let us recall how the natural substances born of starch come into being. We saw how starchy matter is etherealized in the ex-haling of the cosmic breath and changes into sugars, blossom colours, scents, honey, etheric oils and therapeutic substances of plant origin. Downwards from the middle zone of starch there is a gradual densifying and mineralizing, via cellulose, until a biological zero is reached in carbon, or coal tar. Here human ingenuity takes hold and conjures forth a synthetic mirror-image of the natural world: synthetic colours, scents, saccharine and other sweeteners, mineral oils and therapeutic substances.

Contrasting the two realms, we get the impression that the upper one is the realm of dynamic biological reality, the scene of a ceaseless harmonization of the living polarities of earth and heaven, giving rise to an endless range of metamorphoses. The underworld of coal tar chemistry, on the other hand, seems – figuratively speaking – like a ghostly reflection of the dynamic creativity of the cosmos. Here, however, the static world of atoms and calculable happenings takes the place of dynamics. Despite the calculable certainties found in this sub-earthly realm, it cannot seem more real to us than that of the greening, flowering and fruiting plants.

CHAPTER SEVENTEEN

High Dilutions and Their Effectiveness

When a substance such as table salt is dissolved in water, its physical shape disappears completely. It is no longer either visible or tangible. The tongue is the only sense-organ still able to perceive it, as a taste. What has happened to the substance? Is it still there, and if so, in what form?

We say that salt is 'in solution'. We can taste it and demonstrate its presence chemically by combining it with a silver nitrate solution. This precipitates a thick white cheese-like substance, silver chloride, proving that a chlorine reagent was present in the salt of the original solution. A reaction has occurred in the watery fluid, causing a precipitation of silver chloride, a visible indication of the presence of the invisible table-salt.

$$NaCl + AgNo_3 \rightarrow AgCl + NaNO_3$$
Table salt   Silver Silver chloride   Sodium nitrate
(Sodium chloride) nitrate (precipitates) (remains in solution)

But what form does the dissolved salt take?

When a salt solution and pure water have been allowed to stand awhile in a cylinder where they are kept separate by a specially prepared piece of parchment, an upward bulge appears, caused by what seems to be strong pressure from below.

This phenomenon is called osmotic pressure. The prepared parchment forms a semi-permeable wall which lets water through but keeps out other substances. Research shows that the pressure depends on the concentration of dissolved matter, and, further, that this matter behaves like a gas. Thus osmotic pressure can be equated with the pressure of gas shut up in a container.

With ever greater dilution, the dissolved substance keeps on expanding like a gas, finally becoming so dilute that it cannot be detected either by taste or by chemical reaction. Even the subtlest means of chemical detection, such as can determine the presence of substances in comparatively high dilution, gradually fail. Even the very sensitive method of spectral analysis has its limits.

With a dilution of $1:10^{17}$ a point is reached where, from the standpoint of present conceptions of the size of atoms, one would have to say that not a single molecule of salt can still be found in 1/1000 ml. of the solution. (According to Loschmitt, 1/1000 ml. of gas contains 4.5×10^{16} molecules.) This is the same as dissolving 1 mg. of table salt in ten million tons of water. With further dilution, even a larger amount of the solution will not, theoretically speaking, contain any salt molecules at all. It is therefore hard for chemists and physicians with a modern scientific training to put any faith in the effectiveness of high dilutions.

Nevertheless, a steadily developing homoeopathy has been demonstrating for more than a century that these and even higher dilutions have therapeutic effects.

Hahnemann was the first to make therapeutic use of high dilutions. His method still governs the preparation of drugs used in modern homoeopathic practice. Dilutions made by Hahnemann's method are called potencies. They are prepared in the following way:

A gramme of some substance, copper sulphate for example, is dissolved in nine grammes of water, making ten grammes of a 10% copper sulphate solution. The solution is then rhythmically shaken for a few minutes. This is called the first potency. Recent research has discovered the optimal length of shaking time as well as the most effective shaking rhythm.

Now 1 ccm. of the solution thus obtained is again diluted with water to make 10 ccm. and again rhythmically shaken. The resulting fluid is the second potency. Again, 1 ccm. of this is mixed with nine parts of water and shaken. So we arrive at the third potency.

Since we are dealing here with dilutions of one in ten, pharmacology calls them decimal potencies, abbreviating the term to 'D' with the number of the potency attached. Thus D3 indicates (in German usage) the third potency. At this point the original substance has reached a dilution of 10^{-3}.

From the modern chemist's standpoint the so-called third potency is, of course, just a watery 0·1% copper sulphate solution. In his view it would be simpler to make it by dissolving one gramme of copper sulphate in a litre of water.

But the fact is that a D3 copper sulphate potency is not the same thing as a simple 0·1% solution. This can be proved experimentally. Biological tests were carried out on seedlings growing side by side in potentized and non-potentized equal dilutions (Kolisko). The spread of leaves and roots in the two categories is quite different, and repeated testing always confirmed the difference. A new force is at work in potentized substances, a force aroused by rhythm. With increasing dilution the substance disappears, first physically, then chemically and spectroscopically, while biological activity is increasingly enhanced.

If we leave aside the contemporary concept of substance, bound up with the idea that matter is everlasting, and take instead the new dynamic concept set forth here, the problem of high dilutions is immediately clarified. As we have shown above, matter is a crystallization of macrocosmic processes. What we call substance on earth is cosmic process in a fixed, rigid form. Earthly materiality and cosmic being are the two poles between which nature unfolds her myriad manifestations. The plant is placed in the midst of this polarity, a living member of the world organism, subject to countless metamorphoses of form and substance, involved in rhythms of contraction and expansion, involution and evolution, being and appearance.

Our experiments have shown how matter appears and disappears in cosmic rhythms. Graphs traced the constant oscillation between material condensation and etherealization into an imponderable state of being (cf. figs. 1, 2, 3 and 5). Goethe described this rhythmic process in relation to the plant. We can extend his concept to include all substances. In every case, a being, a macrocosmic idea, is rhythmically metamorphosed into what we call matter. And these fixed stages of substantiality can again be rhythmically brought back to the original form of essential process, like plants withering at the approach of autumn and withdrawing almost wholly from the physical scene as their essential being returns to the wide reaches of the universe.

The potentizing method is an inspired emulation of this natural process. It is simply a conversion of matter from appearance to being; the substantial is etherealized by rhythmical dilution. Physical shape disappears as the substance changes from a solid to a liquid, then to a gas, and finally passes over into a still more refined, non-material condition.

Homoeopathy aims at healing effects achieved not by resort to matter in a crude, fixed state, but by the release of what we might call the free, active spirit of each given substance.

It seems that Hahnemann indicated the ten-rhythm used in potentizing. This has continued to govern practice ever since, as no new creative ideas have appeared to change the satisfactory old-established ways and because the ten-rhythm is not at odds with the decimal system everywhere in use. But we may reasonably assume that every substance has its own individual rhythm. A clue to such rhythms may be found in the relation of a substance to the formative forces, expressed in crystallization and other patternings, or to hydrogen, as shown in the figures for relative atomic weight (compare these with the musical qualities of the various substances). Or we can note that a plant – mistletoe, for example – is strikingly dichotomous. This might point to using a two-rhythm in its therapeutic potentizing. Such a rhythm has been in clinical use for years in the case of mistletoe, and has proved very effective.

It remains for research in the coming decades to make some discoveries at present undreamt of in this field. The first step is to undertake a study of what are called potency curves.

The experimental method consists in planting wheat grain in thirty flower pots, watering each pot with one of a continuous series of potencies of a given substance, and letting the grains germinate and grow. After a few weeks the seedlings are measured, and the average height reached with each potency is calculated and entered on the vertical arm of the graph (Kolisko method).

These potency curves show a constant characteristic for each substance tested. In the author's laboratory the potency curves for a large number of mineral and vegetable substances have been established.

One change was made in the experimental method, however, in that yeast was substituted for the wheat seedlings. Yeast is known to break sugar down into alcohol and carbonic acid. The latter can be collected in a graduated tube for measuring. The

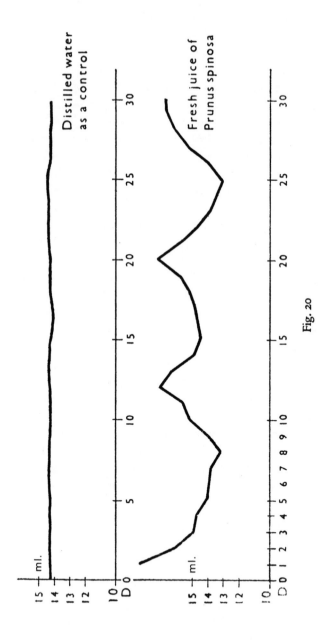

Fig. 20

Potency curve of *Prunus spinosa* (sloe-blossom).

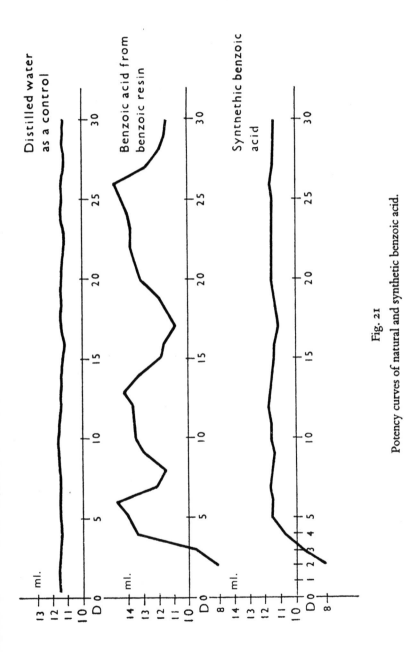

Fig. 21

Potency curves of natural and synthetic benzoic acid.

volume of carbonic acid is as accurate an indicator of the life-intensity of the yeast as were the measurements of leaf and root growth in the case of the seedlings.

The complete apparatus consists of thirty to fifty fermentation flasks, placed in the order of potencies on a round, revolving table. Before each test, a control experiment is run with distilled water. If the same amount of water and yeast are used in each flask, the same volume of carbonic acid can be expected to develop in each, with a straight graph as the result. But when the various potencies of a given substance are used instead of distilled water, we get a characteristic curve with clearly marked maxima and minima in place of the straight line. Fig. 20 shows such a curve, obtained with *Prunus spinosa*.

We have to ask what the dips and rises, the minima and maxima of these curves, really indicate, and whether conclusions as to the rhythm of a particular substance can be read from the intervals between them. And does the curve give any clue to therapeutic use and dosage?

It may take years of diligent research to answer these questions. For the time being we can perhaps say only that the maxima and minima are turning-points, a crossing of thresholds between two realms. We can picture a series of thresholds to be crossed with increasingly high potencies as a substance is brought from one stage of existence to another. The threshold-points (the maxima and minima of a particular curve) may express this ascent of a substance through various phases.

In the course of the author's experimenting, benzoic acid was tested and its potency curve determined. This substance was derived from benzoic resins and purified by being crystallized until the purity standard was reached at the melting point of 121 °C. The potency curve was lively, with characteristic maxima and minima.

A sample of benzoic acid was bought at the same time from the pharmacy and a potency curve test was run. This sample was a pure but synthetic coal tar product. Though it was chemically indistinguishable from natural benzoic acid, it belonged to the mirror-realm of coal tar chemistry.

The potency curve of this synthetic substance did indeed turn out to be a straight line (cf. fig. 21). This means that its higher potentizations had no influence on the yeast culture. Only where there was still some material trace of the substance (up to about the fourth potency), was any effect observable. We may conclude that synthetics of coal tar origin are effective only in allopathic dosage; in homoeopathic dilutions they are useless.

Coal tar derivatives cannot, in our terms, be etherealized to the 'being' stage by a rhythmical potentizing process. They have forfeited all relationship to cosmic and terrestrial rhythms and do not respond to a rhythmic challenge. Thus it is fair to say that a basic biological difference exists between natural and synthetic products, despite their chemical identity; the benzoic acid experiments cited show this.

The two benzoic acids are offspring of two realms, shown in the substance spectrum of the preceding chapter. These realms are governed by different types of laws. The living plant belongs with its whole spectrum of substances to the realm governed by the laws of life: the overarching principles of polarity, intensification and metamorphosis discovered by Goethe are operative there. In the second realm the laws of physical atomic and molecular chemistry prevail. In one case we are dealing with organisms, in the other with mechanisms. Hence the law of the conservation of matter is entirely valid in the mechanical realm, but not in the organic.

In daily life we are surrounded by the products of tar chemistry and must form some conception of what we are dealing with. Nothing that has been said here is meant as a disparagement. But we need to recognize what forces are involved in this current phase of evolution as we work towards a future understanding of living substances.

CHAPTER EIGHTEEN

Minerals

When plants are burned, they leave mineral residues in the form of ash. The amount of ash varies considerably with the species and with growing conditions, and the ashy components vary too from plant to plant. Potassium, phosphorus, calcium, silica, magnesia and sulphur are almost always predominant. Small amounts of aluminium, sodium and chlorine are also usually present. Some plants contain rarer elements; there is lithium in tobacco, iodine in seaweed and lichen, titanium in roses.

Nowadays, people tend to think of these minerals as essential nutrients taken up out of the soil by plants. If that were true, crops would gradually deplete the soil of these elements except where it can replace them of itself. This would require adding minerals to keep land in production. And just such a view gained acceptance toward the end of the last century as a result of the great triumphs of natural science, and of chemistry in particular. It found support in Liebig's balance-sheet of plant growth. He is the father of artificial fertilizers.

Though the theory is logical enough, doubts as to its validity arise on closer study of the principle of artificial fertilization.

At first, mineral fertilization certainly increases yield. But already today, after only a few decades' use of artificial fertilizers, there is room for very grave misgivings – and a few decades are just a moment in the long evolution of our cultivated soils. We must take into account what happens to the quality and health of plants so forced, and weigh their vulnerability to the ever-increasing numbers and species of pests and parasites. Indeed, the pesticide industry has grown in exact proportion to the increasing use of commercial fertilizers. This kind of agriculture grows out of quite other roots than that older form of husbandry which provided wholesome food in the past and can continue to do so in the future.

The mechanical balancing of plant nutrients must be recognized as wholly foreign to the organic nature of the plant. The fact that tilandsia grows on telegraph wires without any contact with the soil, or that tobacco plants contain lithium even when grown on land completely lacking in it, would seem to indicate that a plant's mineral content is not the result simply of absorbing soil minerals but of other processes that have either gone unrecognized or have been ignored.

Von Herzeele's experiments and those of the author, mentioned in preceding chapters, show the invalidity of the mineral-balance theory applied to the organic kingdoms. Life overrides the law of the conservation of matter. We showed how matter is macrocosmic process in fixed earthly form. Is it so surprising that 'heavenly forces pass their golden vessels back and forth', as Goethe put it, changing the nature of their materiality in accordance with the stage passed through in rising to the condition of immaterial force, or again descending from macrocosmic being into solid matter that can be weighed and analysed? Is it not perfectly comprehensible that mineral substances should emerge in plants, generated out of the cosmic whole that formed them and of which they are a part, and that we should find these minerals afterwards in analysed remains of plants or as part of the earth where the plants once lived? The earth does not manufacture plants by some physico-chemical process: it is the plant that creates the soil by coming into material manifestation out of the universe (Herzeele). For the plant, earth forces are only one pole, the pole that enables it to make a material appearance.

Plant physiologists and farmers alike should devote themselves to a study of the organic connection of these earth-forces with the forces of the cosmos and put what is learned into enlightened modern practice.

We may say, then, that the plant produces soil. Not only does it build up a fertile layer of humus over the underlying rocky earth in the course of its long creative life-activity; it makes 'earth' in a much broader sense than would at first appear. We are accustomed to recognize seams and deposits of coal as remains of a long-dead vegetation. But living plants can be recognized as playing an even greater role in the formation of the earth. Of course, neither the earth nor plants were always as densified as they are today. Peat moors can serve as reminders of a process that must have gone on under quite different climatic conditions and on a very different scale in earlier times. Primeval plants can be conceived as first expressing their essential nature in the most delicate and fleeting use of highly rarefied elements, like colours in a painting. The material precipitates of these manifestations grew more and more densified and became the substances that formed the earth's surface. This applies equally to the silica, alumina and other substances of which mountains are composed.

Something quite similar happened in the animal kingdom. There are great mountain areas made of animal remains (mussels, snails – ammonite chalk). This is easy to understand, just as coal is. But animals must be pictured as involved in the same development as plants, the substance of present mountains being a densified precipitate of their primeval life in a still quite ethereal element. In this connection we recall what was described above when we advanced the concept of the pre-existence of spirit: that life was present in essence before becoming fixed in earthly forms and solid substances. The beings of nature's various living kingdoms only gradually create themselves bodies in the course of a long densifying process through which they descend into substance-forming manifestation.

CHAPTER NINETEEN

Limestone and Silica

A person travelling eastward to Austria over the Arlberg Pass into the Rosanna and Inn River valleys and looking at the landscape on both sides of the railway can make some very interesting observations. On the right is an ascending line of mountains patched with woods and meadows. Lush pastures with peacefully grazing herds of cattle stretch right up to the edges of the glaciers. Here is a realm abundantly supplied with water, of which the traveller catches constant glimpses in water-falls, gorges, and gay little brooks leaping down through the meadows.

A lover of minerals would find a stay here most rewarding, for the heights abound in glorious silica rocks and crystals: all the quartzes, including milk and rose quartz, felspar in every form, even garnet-like pieces, and mica gleaming in the sun like polished metal. This mountain range is built of a composite rock, granite; its components, quartz, felspar and mica, are also present separately. The geologist notes differences of structural origin in the granite's quality, ranging from the granular to an almost slate-like foliation. He is aware that the rock owes its form and firm robustness to the quartz in it, while when mica predominates, it tends to split the rock into separate leaves. Quartz and mica are in this sense opposite poles, seemingly harmonized and held together by the felspar in granite.

Just as granite is made up of three different minerals, felspar consists of three different chemical elements. It is a kind of smaller recapitulation of granite's threefold nature, in a sense its heart. Felspar is essentially always a combination of silica and chalky matter (limestone or alkalis) linked by alumina. This can be represented in a general way as follows:

Granite: Mica————Felspar————Quartz
$$CaO.Al_2O_3.SiO_2$$
Calcium—Alumina—Silica

With any luck one can also find silica here in its purest, noblest form: rock-crystal. This stone, with its transparent hexagonal columns topped by compact pyramids that opalesce under light in all the colours of the rainbow, is indeed the monarch of these mountains. Here we see a miniature embodiment of the majesty, the lofty clarity, of the surrounding giants, suffused with primeval grandeur to the furthest reaches of their eternal ice, their glittering snow-fields and peaks reaching up to the heavens. Like them, rock-crystal seems to extend out into the realm of universal silence.

The mountains to the south, or on the right of the traveller, are all built of silica. They stretch from the Verwall and Silvretta ranges south of the Arlberg to the Oetztaler, Stubaier and Zillertaler Alps and even further eastward. Geology recognizes siliceous rock as the earth's oldest, and therefore calls it (in German) 'primal stone'. These mountains form part of the primeval Central Alps. They are the backbone of the Alpine massif, extending in a great arc from the Mediterranean to the Hungarian Plain.

Now let us turn to the left and look at the mountains close by the Arlberg to the north, the Allgaeuer and Lechtaler Alps, and, a little further on, the Karwendel range with its fierce chasms and ravines. One has to be a rapid observer in order to take in all the quite different impressions. No trace of lofty clarity remains. Instead, one is charmed and repeatedly surprised by the impulsiveness and spontaneity of the scene. The earth tends to thrust up abruptly in steep and clefted walls, sharp peaks and projections. Only the foothills are clothed in forests, from which naked cliffs rise directly skyward. This is the homeland of the nervous chamois, not of the phlegmatically ruminating cow. The entire range seems caught up in a desiccating process, and one sees fewer brooks, streams and waterfalls than on the opposite, siliceous side. The rock here is limestone – porous, absorbent, eaten away and hollowed out by wind and weather. There are the dense Karwendel limestone, shell limestone, ammonite chalk, and now and then crystalline calcite or marble.

One might call the primeval mountains vegetative, plant-related, while limestone ranges are skeletal, dried-out, restless and animal-related. In line with the previous chapter we can attribute primeval rock to a vegetative origin, limestone to animal origin. But we should not picture these plants and animals as built of substances like those found in species of the present. They were, rather, precipitates of an ethereal substantiality more like light, or air, or even fire.

The views to left and right of the train are so strikingly different that the traveller can hardly escape the challenge they present to thought and feeling. The majesty of the primeval mountains on the right makes the same impression as wise lofty thinking, while the impulsive juttings-up that typify the limestone ranges on the left seem embodied thrusts of will. Many readers will be able to call to mind such impressions from their own experience of similar landscapes. They will have felt the lift in mind and body given by silica, while a limestone soil, though it stimulates the will to action, keeps one firmly anchored in the here and now.

In this connection it is interesting to note which metals favour limestone and silica as matrix. Gold is found in the primeval ranges of the Gasteiner-Ache, in the Stubai Valley, along with copper, mercury and iron, while the limestone harbours either silver or silver-bearing galenite. One comes across many worked-out silver mines in the Karwendel Mountains.

A reminder that the 'primeval' Alps form the backbone of the great Alpine arc may serve to throw light on the overall lime-stone-silica picture. North of this backbone lie the northern limestone ranges. The valleys of the Inn, the Salzach and the Enns are the line of demarcation here. To the south we again find limestone ranges: the Dolomites, Karawanks, Julian Alps and Karst. Here, the boundary between primeval rock and southern limestone is the Drava River.

Many European ranges are primeval rock, particularly the Black Forest, the Bohemian Forest, the Sudeten and the Scandinavian mountains. Geologists estimate that 40% to 50% of the earth's crust consists of silica, about 30% of limestone. Thus we may properly call limestone and silica the building stones of which the earth is made.

Silica has an affinity to water, as may be seen in the many water courses of the primeval Alps and in the vegetative forces already noted. This affinity shows up chemically as well. No other substance forms so many different compounds, distinguishable solely by their water content. Silica is an acid, an anhydride (silicic acid – SiO_2). In combination with water it changes into orthosilicic acid (H_4SiO_4). In between is metasilicic acid (H_2SiO_3). But these are simply outstanding types. There are innumerable silicic (polysilicic) acids, whose salts (silicates) account for the tremendous variety of silicate rock. We showed above how carbon, in the unique ability to combine with itself which its function as a shaping force confers, creates the vast range of organic substances. Silica has a similar capacity in the realm of minerals, where its affinity to water enables it to create the tremendous variety of compounds composing the various kinds of rock.

Silica's affinity to water is actually so marked that solid particles of it will mingle with water (physically, not chemically) so completely as to give the appearance of dissolving in it. This makes a mixture known as waterglass. It is not, however, a true solution, such as table salt makes when dissolved in water. For table salt in solution not only loses its solid crystalline identity: it permeates the solution as a gas would, and behaves like a gas in all other respects. But silica in solution is at an in-between stage; it cannot make up its mind to be either a true solid or a liquid, or yet a genuine solution. This condition is familiar from our discussions of starch and protein; it is known as colloidal. It is a labile state that can change in a twinkling, going over into a solid by way of gelatinous (hydrogel) and flaky stages, or becoming fully liquefied (hydrosol) and turning into a true solution.

It is most interesting and revealing to look at the picture colloidal chemistry gives of the reversal of structure that takes place as a hydrosol changes into a hydrogel. Solid particles of silica are assumed to be suspended in the fluid hydrosol. When this passes over into the gelatinous condition of the hydrogel, the structure is reversed. Where we had solid silica particles in water, we now have liquid-containing hollows in a solid mass of silica very like a bath-sponge in appearance. Opals, for example, are simply hardened silica jelly.

The colloidal state is characterized chiefly by a considerable increase in surface tension. Every substance is densest at its surface, where it has a sort of skin. That is how surface tension manifests itself. Thus if a knitting needle is carefully put down on water it will float there, supported by this 'skin'. In colloids, surface tension is enormously increased, since there is not only the surface formed at the border between air and liquid, but, in addition, all the numberless surfaces of floating particles of silica, or of the drops of liquid in the holes that honeycomb the solid cake. Silica-gel is thus a structure made up of skins and internal energy. All colloids are in this sense energy carriers. They have a maximum of reactive surfaces, always the essential factor in biological processes. As carriers of life, all the fluids in human, animal and vegetable organs are colloidal.

Silica's proclivity to the gelatinous state and its even more pronounced tendency to surface tension give us a clue to the manner of its natural and cosmic functioning. In hydrogel, where we see silica in action as a peripheral force working from outside to bound the hollow spaces, we have a miniature picture of its most outstanding feature. Everywhere we find it forming sheaths and enclosures with its surfaces. It is known that skin and all the surface sense organs that connect us with the world outside are largely composed of silica, as is the umbilicus. Here we see silica again in the role of matrix, transmitting the mother's life and formative forces to the embryo.

Even in the mineral kingdom silica exhibits the same activity. Agate has a spherical structure, while chalcedony looks exactly like pig-skin. Both minerals are pure silica, and their skin-forming tendency is clearly apparent. Looking at them, one has the impression that they got these skins from an outside force that worked in from the cosmos. Nowhere is this more perfectly illustrated than in amethyst geodes. Cosmic formative force, formative light, shapes the skin-like enclosure of these cavities, inside which, as in a womb, the marvellous crystals come into being. And amethyst is purest silica.

To understand these matters in their far-reaching implications we must widen our search beyond physico-chemical limits and observe silica in the entire range of its activity. It is not just an earthly substance; it is a macrocosmic force or process that has shaped the whole earth-globe as a sculptor's hands create form on the surface of some plastic medium. The silica process is a form-giving process.

Projective geometry provides us with a mathematical illustration. One can think of the surface of a sphere as having two possible origins. The usual concept pictures it as an equal expansion from a central growth-point. Every point on the circumference thus has a static connection with the centre. This concept underlies all our building, the statics of a house, the stability of the physical universe.

The alternative method of arriving at the surface of a sphere is just the opposite. Form-creating planes approach the emergent shape from infinity, becoming tangents of the sphere thus enclosed. Such spheres are therefore infinity-created hollow spaces, not solid bodies. Every point on the surface is linked with infinity, rather than with a centre (fig. 22).

This shaping with the cosmic dynamics of infinity is the silica process, which manifests in a fixed form as the substance, silica. It is the force active wherever surfaces come into being: the surfaces of ocean waves and mountains, the epidermis of plants, the skins of man and animals, the membranes enclosing their internal organs.

Silica serves as an excellent example for pointing out the difference between substances as the dead mineral end-products of a process and that process itself – a distinction we shall have to learn to make. The silica process is by no means bound to the substance silica. Only where its activity has been intense, does material silica come into being. Substances are the final stage of processes. Future references to 'substance processes' are to be understood as meaning the dynamic activity preceding the emergence of physical matter.

Birds are the creatures most closely related to the silica process. Material evidence of this is found in their feathers, ashes of which yield up to 77% silica. The silica process here is not limited to the feather-enclosed bird form; it is the force which relates birds to the entire sphere-shaped mantle of air that encloses the earth. Atmospheric strata are themselves permeated by the silica process, for their surfaces were formed by its tangential planes. It is not just strength of muscle that enables the eagle to spread his wings

THE FORMING OF THE CIRCLE:

from the centre –
STATIC (LIME)

from the circumference
DYNAMIC (SILICA)

Fig. 22
Earthly circle and cosmic circle.

and lord it over the heights in majestic flight: it is the profound relation of his being to the silica process active in the airy realm. Gliding is a first step in the same direction, a sport in which man has to develop a special qualitative sense for and sensitivity to the cosmic laws of atmospheric space. It is a sense that borders on perception of imponderable reality. An understanding of the silica process could contribute toward a future mastery of the air, which is structured throughout of plane-surfaces.

How different a picture limestone gives! Here we find no such affinity to water as silica possesses, but rather an inclination to aridity. There is no such thing as colloidal lime. Our path to an understanding of its nature cannot therefore trace a relationship to water, as with silica, but rather a connection with dryness.

Now the greatest intensification of which dryness is capable is combustion. Firing produces what we call burnt lime, but leaves silica completely unchanged. Lime in its natural state has a greediness, evident in its tendency to absorb liquids, gases and odours. Burnt lime carries this characteristic to an extreme. In the process called slaking, for example, it sucks up water with such ferocity as to cause hissings, clouds of steam and even explosions, so that it has to be very cautiously handled in the lime-pits. Slaked lime is made by pouring water on to quicklime. And though one might think this would satisfy its thirst, it goes on and greedily sucks up more carbonic acid from the air, until it becomes hardened stone again. That is why lime is an ingredient of the mortar (a mixture of sand and slaked lime) used in construction work.

Burning: $CaCO_3$ (lime) $\rightarrow CaO$ (burnt lime) $+ CO_2$
Slaking: CaO (burnt lime) $+ H_2O \rightarrow Ca(OH)_2$ (slaked lime)
Building: $Ca(OH)_2$ (slaked lime) $+ CO_2 \rightarrow CaCO_3$ (limestone)

Lime is thus related by its character to the statics of building, the firmness of our physical frames, the equilibrium and stability of earthly phenomena.

Fig. 22 pictures the contrast between the dynamics of the cosmic sphere and the statics of the earthly globe. The latter is built up from a central point and obeys the laws of terrestrial space, while the former is shaped by the peripheral forces of infinity.

We seldom think of the extent to which as earthly beings we are bound up with the balance and stability derived from the lime-process. Gravity would drag us down into bottomless

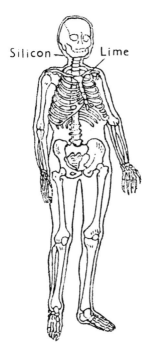

Silicon — Lime

Fig. 23

Skin and skeleton as expression of the silica-lime processes.

depths if the earth's solidity and the statics of buildings were not there to counteract it and to hold things in balance. And we can thank the lime in our bones for the firm support they give to our bodies, for our balance and our upright posture.

Our skin, then, is that part of us built by the cosmic silica process, while our skeleton is the expression of the stable earthly minerality given us by the lime-process. With our 'skin and bones' we are stretched between the two polar cosmic forces, silica and lime. Silica shapes from the circumference, lime from the centre. The dynamic relationship of the two forces is essentially that of circle and radius.

The silica-lime polarity finds reflection in another fact. From the chemist's standpoint, silica is an acid, lime a base. Bases cause precipitation, while acids are dissolving agents (metals and minerals cannot be dissolved directly by silicic acid, but they can be by almost every other acid). Precipitation occurs in almost every case where a base is added, thus actually creating a 'basis'. Such is the origin of the solid ground that provides the basis for our feet, our lives, our work. The acids and bases recognized by chemistry to be polar opposites are characteristic expressions of silica and lime, which are themselves expressions of polar forces of the universe.

Where in the universe is the origin of these macrocosmic silica and lime processes? We must seek it, as we have done in other similar cases, in the realm of fixed stars, the Zodiac.

As we have shown, the silica process shapes life out of the cosmos as a sculptor's hands shape clay. But hands are simply tools for working out ideas that live in the mind of the artist. Macrocosmic ideas underlying outer forms of life: this is the heart of the silica process. They are what we might call archetypal images or pictures. Goethe referred to them in Faust as 'The Mothers'. The sun, which is the mediator of these forces, transits the constellation of the Ram in April, just that time of year when nature rises to bring forth a wealth of new forms – in other words, the season when out of their abundance the archetypes are manifest in physical appearance. Substances more resistant than clay cannot be smoothed and kneaded into form; they must be hammered into shape. The ancients expressed this great process, whereby the world of appearances selflessly receives the imprint of archetypal patterns from creative heights, in the symbol of the backward-glancing Ram.

Aries, the Ram (also the Lamb) is a wonderfully profound image of the cosmic process which gives rise to silica as its last mineralized stage. As with our earlier picture of the Lion, so now the activity that proceeds from the Aries region of the heavens is pictured at every level in the Ram. For one thing, the ram's twisted hollow horn-formation is a pure example of the characteristic sheath-building, sphere-forming activity of silica. This spherical tendency can be seen also in the ram's whole body-surface, in the wool that covers even its head. This soft, bright, silky fleece was from time immemorial a symbol of high cosmic powers, of wisdom that prevails in the realm of archetypes. All genuine early representations of the Ram show him looking backwards, emphasizing the way in which the silica process rays in from universal space.

The autumnal equinox, when the sun is travelling through the constellation of the Scales, ushers in a season which is just the opposite of spring. Whereas in spring the atmosphere is vibrant with the urgency of growth taking shape, by autumn this wave of materialization has spent itself. An ebbing tide sets in; the world of archetypal images withdraws. Earth's static tendency regains the upper hand. The Scales thus accurately picture the lime-forces and their point of origin.

As earthly substances, silica and lime are polar opposites. Their macrocosmic archetypes are similarly polar, for the Ram and the Scales confront each other from opposite sides of the Zodiac.

CHAPTER TWENTY

Aluminium and Phosphorus

Valleys and the basins between mountain ranges delight us with their luxuriant vegetation. Here we find still another type of landscape. Everyone knows the kind of soil native to such regions – its brownish-yellow clay sticks to one's shoes in rainy weather and one's feet sink into it. Chemically speaking, this loam is an aluminium silicate. In its pure form it is known as potter's earth or china clay. This is the type of soil deposited in river valleys or in mountain basins. It is a heavy, fertile soil, such as favours the growth of lush foliage. Geologists call it alluvial because it is formed by a weathering and washing down of disintegrated mountain rock. Thus we may call clay a bridge between ranges. Since it has been found to constitute about 20% of the earth's crust, it is – like lime and silica – one of the building materials of which the earth's body is composed.

Lime and silica were shown to be polarities. Alumina seems designed to bridge the two. We can see this tendency in rock where clay is present. Felspar is a good example. It plays a harmonizing role between quartz and mica, a fact apparent in the chemical formula. According to whether the aluminium in felspar is combined with calcium, sodium or potassium, it oscillates between silica and lime compounds:

Felspar:	CaO	.	Al_2O_3	.	SiO_2
(type)	Calcium		Aluminium oxide		Silica

This pendulum-swing between polar opposites becomes even more evident on considering aluminium's 'amphoteric' nature. This means that it can be both an acid, like silica, and a base, like lime. Taking one of the commonest forms of aluminium, alum, or aluminium sulphate, we find that in this compound aluminium acts as a base, combined with sulphuric acid. When alum is dissolved in water to which a stronger base such as caustic soda is added, aluminium hydroxide is precipitated:

$$Al_2(SO_4)_3 + 6NaOH \rightarrow \qquad 2Al(OH)_3 \qquad + 3Na_2SO_4$$

Alum   Aluminium hydroxide

An overdose of caustic soda, however, has the surprising effect of re-dissolving the aluminium hydroxide precipitate. In the presence of this stronger base, aluminium makes a right-about-face and acts like an acid in combining with the caustic soda to form a soluble salt, sodium aluminate. Here the part played by aluminium complements the part it played in alum:

$$Al(OH)_3 \qquad + 3NaOH \rightarrow \qquad Na_3AlO_3 \qquad + H_2O$$

Aluminium hydroxide   Sodium aluminate

If a stronger acid, such as sulphuric acid, is now carefully mixed with the sodium aluminate, this will cause aluminium to return to its base condition and separate out again as aluminium hydroxide:

$$2Na_3AlO_3 + 3H_2SO_4 \rightarrow 2Al(OH)_3 + 3Na_2SO_4$$

Sodium   Aluminium

aluminate   hydroxide

A slight superfluity of sulphuric acid again dissolves the aluminium hydroxide, producing the readily soluble alum, or aluminium sulphate. In this salt we find aluminium completely restored to its original condition as a base:

$$2Al(OH)_3 \qquad + 3H_2SO_4 \rightarrow Al_2(SO_4)_3 + 6H_2O$$

Aluminium hydroxide   Alum

So aluminium oscillates continually between the poles of base and acid, linking them in a constantly recreated balance.

We even find the plant's two poles, root and blossom, brought into harmony by the aluminium element. It is not, however, the substance aluminium, but the aluminium process that carries the earth forces upwards from the root, and the sun and stellar forces of the blossom region downwards to the root. Aluminium's material presence in the soil stimulates the plant to this activity. Silica is responsible for colour, scent and finely articulated form, while lime sees to the material filling-out of vegetation from below. Aluminium keeps these earthly and heavenly forces in a living balance. That is why we called luxuriance of foliage, the green middle zone of the plant kingdom, a sure indication of a clayey soil.

Clay is plastic and responsive to formative forces working on it from outside. Just as a musical instrument responds to a musician so plastic clay is the instrument for the music of forms composed by a sculptor.

It is the aluminium process that makes earth receptive to the cosmic shaping forces of the silica process which the great artist, Nature, draws from the cosmic periphery. Silica's affinity to water appears again here in relationship to clay, for it is only when clay is properly moist that it is sufficiently plastic to be receptive to the shaping activity of silica.

But the formed clay becomes static in the drying process, while firing makes it almost indistinguishable from lime. Pieces of sculpture, pottery and bricks are all rendered hard, dry and porous in the kiln (pottery is given a skin of silica in the process called glazing). The lime in mortar holds together the bricks in the house-walls which shelter and support our physical life. Again we see aluminium-bearing clay as the balancing agent between silica and lime.

Of course, there are latent polarities in clay. In itself it is the least aristocratic substance; the forms built of it are the most transitory. This is expressed in the Biblical picture of man's transitory physical body formed of clay. We might say, borrowing this picture, that man is built from head to toe out of such a balancing of heavenly and earthly forces as clay affords. But this 'clay' undergoes a stage by stage upward purifying as man refines it in his various organs, reaching a peak in the eye's transparency; here dark earthly matter has been raised to a level where it becomes permeable by the light of spirit.

Clay thus serves also as a gemstone matrix. Gems are the highest stage of mineral matter, perfect expressions of the harmonious interplay of lime and silica, of earthly anchoring and cosmic shaping. Almost every kind of precious stone is made of aluminium oxide or of a compound of aluminium. The family of corunds, rubies, sapphires, consists of pure aluminium oxide. Other gems, such as tourmalines, emeralds, topazes, zircons, contain aluminium compounds.

In precious stones, aluminium lends itself wholly to silica's cosmic shaping forces; in brick it is given off both to dry, static earthly force of lime.

<pre>
Silica   Lime
moist ←————— Aluminium —————→ dry
Gemstones   oxide   Brick
</pre>

Putting a ruby, with its brilliant red, beside a soft blue sapphire brings home the fact that jewels are a synthesis of polarities at the very highest level of which matter is capable.

There is a wonderful gemstone that combines two polar colours in each single crystal. This is the tourmaline, with its complementary green and purple.

Turning to man, whose physiology lies between skin and skeleton, silica and lime, we find an element which as the carrier of physiological processes moves in ceaseless rhythm between polar opposites. This is the blood, which streams out to the periphery of the body and then returns to its innermost core. As it moves toward the skin and the extremities, blood is red; on its return journey, blue. The heart is like a jewelled expression of this active synthesis. Its beating is a rhythmic harmonizing of these poles.

How understandable it seems in the light of these facts to apply aluminium (in the form of aluminium acetate or clay poultices) in treating congestions, inflammations, sprains and bruises. Felspar (orthoclase), externally applied, also helps to harmonize heart action.

In contrast to clay, phosphorus (or phosphate rock) is thinly scattered through the earth's crust, like spices in a cake, instead of filling up whole regions, valleys and basins. Rarely are phosphate deposits sufficiently concentrated to make mining them worth while. This mineral, found chiefly in the form of calcium phosphate or apatite, is much sought after by manufacturers of superphosphate, a well-known artificial fertilizer.

But phosphorus is everywhere in minute quantities. Humus derives it from decaying plants, and plant-ash has it in considerable amounts. Where dead plant-matter piles up in layer on layer, in swamps or on moorlands, decomposition releases an organic phosphorus compound in the form of will o' the wisp (phosphene, PH_3).

If we put a piece of phosphorus on a plate, all sorts of interesting things can be observed. One is how the piece shines with a peculiar greenish glow in the dark. At the same time we notice a strange characteristic odour, exactly like that given off by a shower of electric sparks. This is caused by the forming of ozone. Phosphorus has the same capacity as electricity to condense oxygen in the air into ozone.

$$3O_2 \rightarrow 2O_3$$
three volumes oxygen → two volumes ozone

Finally we see fumes spiralling around the phosphorus. It does not look as though they were generated by it, but more as though they were circling in toward it as a centre, closing in on it. Suddenly the phosphorus ignites with a brilliant white flash of spontaneous combustion, making a spurting, hissing sound, 'ffft!', as it bursts into flame.

Phosphorus, then, shines and pours out light, but is also a condensing agent. Like aluminium, it embraces two polarities, both of which are to be found in the human organism. Our bodies contain a very considerable amount of material phosphorus. Nerves are built of protein high in phosphorus. Indeed, the nervous system as a whole is as clear a revelation of the phosphorus process as the circulatory system is of the aluminium process. Phosphorus flames give light, but are cold. Our nervous system endows us with the cool, clear light of consciousness; but it is also the transmitting agent for the formative impulses that shape the body's plastic organs. This means that without the nervous system there would be no such thing as shaped, substantial human bodies.

The phosphorus process co-operates on the one hand with the silica in our skin, on the other with lime in our bony structure. The skin contains innumerable nerve-endings, which convey impressions of the world around us. Though silica creates skin surfaces, it is the phosphorus process that gives them surface sensitivity. It is to phosphorus that we owe awareness of our bodies and a bodily consciousness of selfhood. The skin, with its nerve inclusions, thus forms a boundary between world and individual.

The cosmic formative forces active in silica are taken over by the phosphorus process, which, with the help of the transmitting nerves, uses them to shape and give firm substantiality to the internal organs (we may recall the beri-beri syndrome, with its typical poverty of nerve-transmitted formative forces and its tendency to tissue dissolution). The dynamic of this densifying process culminates in the bones, where phosphorus combines materially with calcium to form calcium phosphate. Thus the phosphorus process comes to its completion in the density and statics of the bony structure.

We may say, then, that the poles of skin and skeleton are spanned by the phosphorus-nerve process as well as by the aluminium-circulatory process.

The path travelled by phosphorus from the skin to the bones, via the nervous system, can be clearly traced in the pathology of arteriosclerosis. The nerves with their phosphorus content – a substance known to chemistry as nucleo-protein and composed of lecithin and cholesterol – are at an intermediate stage of calcification. In persons tending to sclerosis, the walls of the blood-vessels are coated with cholesterol and similar substances. As the disease progresses, these deposits calcify and bring on hardening of the arteries.

These facts indicate both the cause of sclerosis and the means of preventing it. The cause lies in a one-sided development of the phosphorus process with insufficient circulation to oppose it. The result of too great intellectuality and self-centredness is physical hardening in later life.

Aluminium and phosphorus may thus be called mineral substances swinging pendulum-like between lime and silica. Each does so in its own characteristic way, which is the polar opposite of the other.

Now let us find the macrocosmic place of origin of these processes.

In winter, when the sun has just passed its lowest point, and at the feast of the Magi (January 6) the days have lengthened by a hair's breadth, we begin to get the feeling of having crossed the threshold between autumn and springtime. The sun stands at this point in the sign of Capricorn, the mountain goat. Strange to say, old representations of this animal show it with a fish-tail. How wonderful a picture of the double nature of aluminium! The goat represents its inclination to lime, the hard and dry, and the fish-tail its affinity to water-loving silica. The whole picture symbolizes the triumph of living plasticity over rocky hardness.

Phosphorus is an isolating agent, despite its capacity to span the polarity of lime and silica; for it makes the individual conscious of his separation from the world about him, thus creating the basis for psychological and physical self-awareness. These forces originate in the zodiacal region which the ancients saw as the 'crab' – Cancer. The calcium phosphate shell that isolates the crab from its surroundings pictures the separateness of conscious selfhood. That which begins in the human organism as bodily awareness, as the sense of being a self housed in a body and possessed of independent thinking, only very gradually completes its formative task on the organs, and finally lays hold of the lime process to harden the skeleton. In the crab's case, the same forces are expended on making a material sheath of calcium phosphate.

Capricorn and Cancer are, then, the cosmic regions that engender aluminium and phosphorus. And like these terrestrial substances they are polar opposites, confronting one another across the great circle of the Zodiac.

CHAPTER TWENTY-ONE

The Mineral Cross

Wesaw in a previous chapter that lime and silica are opposite poles of the cosmos, with aluminium and phosphorus, a second polar pair, engaged in a harmonizing pendulum-swing between the first pair. It would be possible to explore many

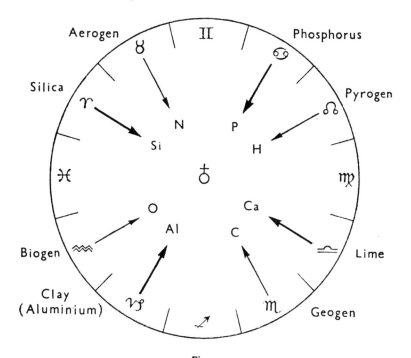

Fig. 24

The mineral-cross (alongside the atmospheric cross).

more aspects of this relationship than there is space for here. The indications given above, however, may perhaps serve to illustrate clearly the harmonious interplay of these four substances, particularly in the field of physiology.

The aluminium and phosphorus processes work, as we have seen, in blood and nerves, linking silica and lime, the skin and skeleton. Blood is the plastic element that builds and maintains the body tissue, whereas nerves transmit the formative forces that take over the plastic material provided by the blood and give it the shape of muscles and organs. The phosphorus process active here is like a sculptor's hand, with its sensitive touch for shaping plastic clay.

Aluminium and phosphorus, then, are a polarity midway between the polar opposites lime and silica. The macrocosmic processes that create these four substances spring from four cosmic points of origin which form a cross: Aries – Libra/Capricorn – Cancer. Just as we found the four substances: hydrogen – oxygen/nitrogen – carbon falling into the pattern of an 'atmospheric cross' which pointed to the cosmic sources of protein and of all organic nature, there is a similar comprehensive significance in the cross formed by silica – lime/aluminium – phosphorus.

It can easily be shown that the whole mineral earth is really built in the main of these four substances. Organic creation with its flowers, trees and grasses, its butterflies, beetles and other creatures is as it were a panorama of fleeting images that seem to emerge out of the atmosphere, take on material form, and just as quickly melt away again. Mineral nature, with its mountain ridges, its plains and valleys, is by contrast the stable, solid core of this passing show. Or perhaps we should say, relatively stable; for it too came into being out of the great reaches of the universe, and will one day dissolve back into it.

In a preceding chapter we related the atmospheric cross formed by hydrogen, nitrogen, oxygen and carbon to the four Aristotelian elements: fire, air, water, earth. This fourfold principle, in which earth's evolutionary phases can be recognized, is so fundamental that we may expect to find it again in the mineral cross. Disregarding for the moment the natural, earthly state of these minerals and considering them purely as principles, or processes, we do indeed find the four Aristotelian elements reflected in them. The relationship of the highly inflammable phosphorus to fire is immediately obvious. Then there is the silica process with its affinity to lightness and buoyancy, exemplified in the flight of birds and the spherical tendency inherent in the formative forces of the macrocosm; here we discern a relationship to light and air. Aluminium, which has so strong a proclivity to the plastic, to the levelling, flowing liquid element, leaves no doubt where it belongs. And the statics of lime clearly shows its relation to the earthy element.

To sum up these conclusions so far:

Aristotle	Atmosphere	Macrocosm	Geosphere	Macrocosm
fire	hydrogen (H)	♌ Leo	phosphorus (P)	♋ Cancer
air	nitrogen (N)	♉ Taurus	silica (Si)	♈ Aries
water	oxygen (O)	♒ Aquarius	aluminium (Al)	♑ Capricorn
earth	carbon (C)	♏ Scorpio	calcium (Ca)	♎ Libra
	Organic nature		*Mineral nature*	

CHAPTER TWENTY-TWO

Alkalis and Halogens

The largest, most inexhaustible source of salt is the ocean, where the percentage of dissolved salt is 3% or more. This salt is extracted by a panning or refining process in which the evaporation surface is increased and the sea water subjected either to concentration or to evaporation. Similar processes have been carried out by nature when, in the course of earth's evolution, a catastrophe or other geological event made an inland sea of some part of the ocean. This is the origin of salt deposits. There are many spots in Europe where such deposits have been found, often enclosed by layers of rock. Nearby towns sometimes got their names from this proximity – for example, Salzburg, Halle* on the Saal, Reichenhall, Hallstatt, Hall in the Tyrol, Leopoldshall, Schweizerhall, and so on. One of the largest salt deposits is at Stassfurt, Anhalt.

Geologists have reckoned that the amount of salt contained in the ocean would be more than enough to construct all the land now above sea level, including mountain ranges. This means that there is as much salt in the sea as there is solid rock on land.

What do we mean when in chemistry we describe something as a salt?

We have already recognized bases and acids to be a polarity. The base and acid tendencies already noted in lime and silica respectively come to clearest expression in the polarity represented by alkalis and halogens. The harmonizing of these polar forces produces salts. Salt is thus a mineralized state of balance between base and acid-forming forces, between alkali and halogen.

To understand this better we must first look into the essential nature of alkalis and halogens. Since salt is composed of both a base and an acid, we can bring out the character of the alkali by eliminating the acid.

There is an old-established chemical process for converting salt into soda, now known as the Leblanc soda process. Salt is heated with sulphuric acid, making sodium sulphate. This is then

* The German 'Hall' means 'salt'.

142

brought to a red-hot glow over a coal fire, which reduces it to sodium sulphide, and treated with lime to yield soda.

$2NaCl$	$\rightarrow$	Na_2SO_4	$\rightarrow$	Na_2S	$\rightarrow$	Na_2CO_3
Sodium		Sodium		Sodium		Sodium
chloride		sulphate		sulphide		carbonate
(salt)						(soda)

Various other processes are currently employed in commercial soda-production, but the one described above and named after Leblanc is the classical one still in use.

Soda is found in nature in the form of double-salt crystals on the shores of salt lakes in Egypt. Egyptians called salt 'neter'; the Latin word for soda, 'nitrum', was derived from it. Alchemists of the Middle Ages borrowed the term when they called salt-petre 'sal nitri'. Later on, chemists called soda 'natrum' to distinguish it from saltpetre, which was termed 'nitrum'. We get our term 'natrium' for the salt from which soda is made from the same source.

Soda comes close to being a true alkali. All we have to do to get pure caustic soda is to 'sharpen up' or causticize a soda solution with slaked lime.

$Na_2CO_3 +$	$Ca(OH)_2$	$\rightarrow$	$2NaOH$	$+$	$CaCO_3$
Soda	Slaked lime		Sodium hydroxide		Calcium carbonate
			(caustic soda)		

Caustic potash (lye) is similarly derived from potassium carbonate, a substance found wherever plant matter has been reduced to ashes. Wood ashes (potash) consist chiefly of potassium carbonate. They have long been used in laundering, and as lye, their caustic form, in soap manufacture.

$K_2CO_3 +$	$Ca(OH)_2$	$\rightarrow$	$2KOH$	$+$	$CaCO_3$
Potash	Slaked lime		Potassium hydroxide		Calcium
			(lye, caustic potash)		carbonate

The 'Stassfurt waste-salts' are the chief source of potassium salts at the present time. When sea water, which contains common salt in a proportion of one hundred parts to approximately two of potassium salts, is evaporated, the lighter potassium salts stay in solution longer than the common salt does. They then form a layer only a few metres thick on top of the salt deposits, which at Stassfurt reach a depth of nine hundred metres. This top layer had to be removed to get at the common salt below – at that time the only part that was considered valuable. This gave the potassium salts their name of waste-salts.

Sodium, potassium, and a few other rarer bases, such as lithium, rubidium and caesium, are all called alkalis, and they are chemically and physically very closely related.

One special characteristic of all the alkalis stands out in the following experiment:

Some metal salt, such as copper sulphate or silver nitrate, is carefully dropped into an alkali solution. The drops do not immediately merge with the solution, but keep to their drop-form, enclosed by delicate, veil-like skins. Often they dissolve slowly into a colloidal system, particularly in the presence of 'protective colloids' such as proteins. Colloidal solutions, as we know, are especially prone to surface tension. One can picture them as liquids with tiny droplets or particles evenly dispersed through them. Every such droplet or particle in these solutions may be described as no longer subject to ordinary earth conditions, for in the colloidal state it has a protective skin that keeps it from combining with other substances. Alkalis, then, have a proclivity to form enclosing sheaths, and the colloidal state may be looked upon as a further development in this direction.

As we know, the body fluids chyle, lymph and blood serum, are colloids, as is the sap that is the life-blood of the vegetable kingdom. All the up-building processes having to do with growth and nutrition in plants, animals and man alike are maintained by alkaline colloids present in the fluids of the various organs. In plants this alkali is chiefly potassium; in men and animals, soda. Up-building processes in man are localized in the area between the intestines, liver and kidneys. Physiologists and doctors are very familiar with the importance of alkalis for the liver functions. Here, where both in man and animal vegetative processes are especially active, we find potassium, the characteristic plant alkali.

The capacity to form enclosing sheaths is the most significant aspect of alkalis, as may be noted in the case of those used in ordinary daily life. We see this capacity very clearly in the cleaning and laundering properties of alkali compounds. They are especially effective in combination with fats and oils. Soap is just such a combination. It is colloid, and produces an emulsion, foam. What is foam other than a great increase in surfaces? Bubbles of foam envelope objects and particles of dirt, and soften them. A woollen cloth dipped in clear water does not always even get wet. But in soapy water every fibre and each least speck of dirt on it is at once fondly embraced, surrounded, softened and dissolved.

We described oil as condensed cosmic warmth. Soap, which is a boiled mixture of oil and alkali, is thus a carrier of enclosing warmth.

The sheath-forming potentiality of alkalis is not to be equated with silica's form-creating surface action. Silica is itself a cosmic sheath, whereas alkalis are simply earthly sheaths that enclose whatever comes into their domain.

If one searches for an appropriate picture to express artistically the nature of alkali's enclosing gesture, we come upon the pictures of a maternal organism giving shelter to the child-to-be. What is meant here can be experienced by contemplating the Sistine Madonna. Here we see the mother surrounded by a host of angels, bearing in her arms a child that seems to be one of them. She has wrapped her mantle protectively around it. We feel in the gesture her deep connection with the heavenly powers for which she has provided earthly shelter.

In ancient times, when no one doubted that the terrestrial is always a housing for the spirit, this truth was felt to be pictured in the constellation of Virgo, the virgin. Here could be experienced the sheath-forming power, whence forces of fertility and ripening rayed down to earth.

The sun, which mediates these forces to the earth, passes through the constellation of Virgo in September, the season when all vegetative burgeoning is concentrated in the swelling fruit. These fruits harbour the future in the form of seed ripening. We might say that an apple can be seen as a picture of brimming sap confined within a form by the Virgo forces active in the alkali.

Now, let us go on from these explorations of salt's base-forming aspect to consider its acid-forming aspect.

Our senses provide immediate clues. Most alkalis are thick, even oily-appearing fluids, especially in concentrated form. But acids, especially the halogens, are usually thin and runny, and in their pure form can even be gaseous. Alkalis taken into the mouth seem to be expanding and filling it, while acids are sour and contractive. Alkalis are slimy and slippery to the touch; diluted acids feel astringent.

One can conclude from these experiences that alkalis, like everything hospitable to life, are of a waxing, flowing, expansive nature, while acids are dry, contractive and hostile to growth. In concentrated form, or on longer exposure, they attack other substances aggressively, searing the skin, for example, and making wounds like burns. They break down, burn, or dissolve what comes in contact with them. They are the solvents used on ores and metals. They have a close bond with hydrogen, enhancing its destructive or dissolving fire-force. Alkalis, in contrast, have an affinity with water, or else with oxygen.

Alkalis, then, are passive, receptive, support-giving; acids, positive and active.

These characteristics show up very clearly in connection with colour. All natural plant-dyes – litmus, or the juice of some berry or fruit such as cranberry, elderberry or cherry, or a blossom colour – move in the presence of an alkali to the passive, dark side of the spectrum: blue, or violet. The addition of one small drop of an even slightly acid substance makes the colour move toward the active, light side and turn yellow, orange, and red.

Alkalis conduce to the colloidal state. They increase surface area, and enfold things, thus helping life to flourish. Acids are hostile to the colloidal condition and the developments it encourages. They press towards decisive action; they either curdle colloids or reduce them to a true solution. Fresh milk is an example of a colloid. Souring curdles it; it separates into curds and the transparent whey.

The acid-forming agents known as halogens – fluorine,

chlorine, bromine, iodine, and their acids, hydrofluoric, hydrochloric, hydrobromic and hydriodic acids – are as like each
other in character as are the alkalis among themselves.

Chemists are familiar with the similar behaviour of these
substances, as is the physicist with their peculiar light reactions.
In an earlier chapter we spoke of iodine as a light-thief. This last
and densest member of the halogen group manifests its relation
to light only in this overwhelming, total – one might almost say
brutal – manner. Fluorine, the first and lightest of the halogens,
exhibits the phenomenon of fluorescence. Its acid, hydrofluoric
acid, is the strongest, chemically speaking. It is such a powerful
solvent that it can dissolve glass. We can melt the broken end of
a glass rod and round it off by using hydrofluoric acid just as
easily as with a bunsen burner.

This capacity for rounding off can be seen at work in the
human organism too, and most clearly in the shaping of the
teeth. When a child's second teeth come through, they sometimes looked jagged and even broken. This is because their surfaces are still unfinished, a condition due to a disturbance of the
fluorine or hydrofluoric acid process. Just as the broken end of a
glass rod can be melted into a rounded edge, the jagged teeth
are rounded off by enamel as they emerge.

Other bodily processes, too, are related to the fluorine process.
There is an illness in which the development of certain extremities such as the nose, chin, fingers and toes is not rounded off
and brought to a close, but continues indefinitely. Here, again,
the defining process which should give every last part and form
the full working through and rounding off proper to it has not
taken hold as it should have done. And the same condition can
develop in the digestive tract when food is not properly metabolized and the contents of the intestines are insufficiently worked
over and formed. We call it diarrhoea, and its cause may also be
sought in a disturbance of the fluorine process. All these cases
can be treated with potentized preparations of fluorspar (calcium
fluoride), to stimulate the fluorine process.

All organic disturbances affect our thought-life in due course,
and the above is no exception. How often we witness an in-
capacity to 'think things through', to draw the necessary con-

clusions from a train of thought! Loss of memory may be its final consequence.

We see the characteristic activity of the halogens in these processes. It is cosmic activity, pressing urgently toward the conclusion of some train of action, rounding our destiny, as often with destructive violence as with constructive, creative impulses.

The ancients saw these processes proceeding from the constellation of Pisces, the fish. This symbol is not readily understood without reference to its ancient meaning. The four last signs of the Zodiac were pictures of human occupations: Sagittarius is the hunter, Capricorn the animal breeder (with the fish-tail picturing the taming of the wild beasts), Aquarius the tiller of the soil, and Pisces the trader on his ocean voyages. The paired fish of Pisces were also a symbol of the feet, and so of travelling. Here we find another indication of the characteristic activity, the moving toward conclusions, the fulfilling of destiny which we have been describing. One does not think out one's destiny; one 'walks' an appointed path. Just as Virgo symbolizes the selfless offering of an enclosing sheath within which other life develops, Pisces pictures an active coming to grips with the world and destiny. And like the alkalis and halogens which they create, Virgo and Pisces are antipodes confronting each other from opposite sides of the cosmos.

Receptive love and ego-like activity pour through the heavenly and earthly spheres from these two macrocosmic points of origin, Virgo and Pisces. But their interaction engenders a third force: the salt of the earth, which in the realm of life stands for the balanced organism; in the realm of the spirit, cosmic evolution.

CHAPTER TWENTY-THREE

Magnesia and Sulphur

Most people know that the sea contains magnesium in the form of salts, particularly magnesium sulphate. Sea salt consists on average of sixteen per cent of magnesium salts. We mentioned the fact that there is enough common salt dissolved in the earth's oceans to build all the continents and mountain ranges. The proportion of magnesium in the salt would suffice for the building of one entire continent. By contrast with such a huge mass of magnesium, the amount of magnesium-bearing rock found in the earth's mountains, in the form of magnesite, dolomite, and the magnesium silicates like mica, hornblende and asbestos, is negligible.

Magnesium salts, however, are found in tremendous layers in the Stassfurt salt deposits. These are the chief workable source of magnesium and its salts on the European continent, but their sheer volume and very limited use make them a troublesome dead load for the potassium industry.

To form a conception of the nature of magnesium we shall have to concern ourselves with magnesia-bearing rock. Magnesite – magnesium carbonate – stands out, for it is used industrially in burnt form. During the firing process it changes into magnesium oxide. It is this oxide's resistance to heat that makes it so valuable in manufacturing. High temperatures fuse an initially light, fluffy powder into rock almost impossible to melt. This capacity to resist heat, which holds up under temperatures as high as 2,000 °C. and more, makes magnesia valuable as a lining material in steel-smelting furnaces. So we may say that it preserves its static character even when attacked by fire, but with a behaviour different from that of lime. While lime becomes violent on firing, magnesia remains calm and gentle; it is not given to hissings, greedy devourings, and corrosive action. Quicklime is a corrosive base, which earns it the name corrosive lime. Magnesia is a mild base.

Magnesia's resistance to heat is coupled with another quality: it radiates light with an intensity hard to equal. When magnesium burns and turns into magnesia it makes a blinding white light that casts shadows even in full sunlight. So we see that the sun's rays themselves, as they reach the earth, cannot match the intensity of light magnesium gives off. This property causes magnesium to be widely used in the making of all sorts of lighting equipment.

This power to ray out, so characteristic of magnesia, also appears morphologically in the ray-formation of magnesium-bearing rock. Magnesium silicates, such as actinolite, serpentine, talc, asbestos, and the like, tend especially to a radiating or fibrous structure, reminiscent – as asbestos is – of textile fibre. Asbestos is actually used to make fireproof thread, cloth, rope and wall-boards.

A further phenomenon is produced by magnesium's affinity to light. Those who have visited the southern Tyrol and witnessed there the glorious spectacle of the 'Alpine glow', will remember its beauty for the rest of their lives. These mountains, the Dolomites, are built of so-called dolomitic limestone, an iso-morphic mixture of calcium carbonate and magnesite. It is harder than ordinary limestone and does not as a rule have the radiating structure usually found in magnesia rock. But when the sun has sunk behind the horizon, the light these peaks have absorbed shines out at the onset of darkness with a gentle rose-red glow.

This unparalleled affinity to light possessed by magnesium explains its presence in the substance chlorophyll and the role it plays in assimilation. Plants, as we know, are made of light, and in the sheaved rays of cellulose fibres we really have, so to speak, materialized sun rays. Here magnesium shows itself in the role of a light-propellant in assimilation. It is magnesium that thrusts light into the dense materiality of starch and cellulose. The same propulsive forces are at work in spring when seeds, which contain considerable amounts of magnesium, begin to germinate, often thrusting up heavy layers of earth or snow in the process.

The same dynamic function serves the human organism wherever solid matter is excreted or separated off from fluids. We see this happening most obviously in the digestive process, at the point where waste matter is separated out of the chyme and takes on a firmer consistency. The drastic effect of ingesting

Epsom salt (magnesium sulphate) indicates the close connection of magnesium with intestinal functioning.

Deep inside the organism there are other eliminative processes which must be recognized as such. One of these is the depositing of bone-building matter in the skeleton, likewise a moulding of solids out of fluids. This process is most obvious in young children, whose organisms are still very plastic and as yet scarcely mineralized. Their bones are subject to a hardening process that reaches its culmination and completion in the emergence of the second teeth, the last and hardest product of the body. At this point the solid organism has been separated from the fluid. And forces that previously served organic functions are now set free, in the form of a capacity to think and remember, making the child ready to begin his schooling.

It is the magnesium process which we see at work in all such developments. On the one hand it plays the role of a hardening agent, compressing life into solid earthly form. On the other it activates light-forces. Thus it combines startlingly contrasting functions.

These contrasts are represented in zodiacal imagery as the centaur. His horse's body symbolizes ties with earthly animality, yet with the rest of his being he raises himself to a luminous human height. In mythology the picture of the centaur with his bow and arrow has always been the symbol of these contrasting forces. The ancients experienced them as proceeding from that part of the heavens known as the constellation of the hunter, Sagittarius.

In sea water along with magnesium, we find sulphur in the form of magnesium sulphate. This gives some idea of the enormous amount of sulphur dissolved in the various seas and oceans.

Free sulphur, which occurs at Girgenti in Sicily and at Murcia and Albacete in Spain, has in the past been attributed to volcanic action. Lately it has been viewed as a product also of the ocean. According to this school of thought, sulphuric acid salts such as anhydrite $(CaSO_4)$ + Kieserite $(MgSO_4.H_2O)$ are subject to a kind of putrefaction which produces hydrogen sulphide. Precipitated in the form of sulphur partly by the air, partly by bacterial action (*Beggiatoa alba*), it works its way into sedimentary layers of clay and limestone and is then extracted by a smelting process.

Volcanic sulphur, which is still being deposited around Mt. Etna, comes from deeper strata in the earth. It would seem that sulphur originates in processes taking place deep within the earth, a finding that need not contradict the theory of putrefying sulphuric acid salts. Sulphur deposits invariably occur in regions heated by volcanic action.

To trace the origin of sulphur is to have described its being: it is a substance with a great affinity to warmth. It is inflammable, burning with a very hot, dark flame. It combines readily with the warmth-bearer, hydrogen, making hydrogen sulphide (H_2S). This is a gas released by putrefactive processes, with the characteristic odour of rotten eggs.

This means, in turn, that there must be some sulphur in protein and in all organic substances that give off hydrogen sulphide in decay. And we do find it in all living organisms, even if sometimes only as a trace. This fact indicates how vital sulphur is to all organic life.

In this connection, one characteristic of sulphur that no other substance exhibits to a like degree is of special interest. It possesses such a capacity for change that we find it in six to seven modifications. We distinguish rhombic, monoclinic and amorphous sulphur, as well as two liquid forms, one a thin fluid and the other a thick fluid known as lambda and mu respectively, and a plastic sulphur like elastic. All these forms can exist within a comparatively narrow range of temperatures. One form can be changed into another simply by heating.

It is very easy to make a colloid of sulphur. According to the size of the particles in sulphur brines, the colour ranges from light yellow to red, purple and blue; in reflected light the solutions are blue or green. According to J. Hoffmann, the ultramarine colour obtained from clay and sodium sulphide also results from the formation of a blue colloidal sulphur. The many industrially important 'sulphur dyes' produced by the Cassella firm are probably also colloidal forms of sulphur. They are made by heating various organic substances with sulphur and alkali sulphides.

Chemically speaking, sulphur is the most active of all substances. It does not act in any one clear direction as do the halogens and oxygen; rather does it form combinations in a sociable way, creates new possibilities and supplies warmth, acting towards other substances as a kind of cook.

Sulphur's function in protein is of just such a nature. Its liking for the colloidal state and thus for everything alive, its capacity for change, its brood-warmth make it a natural mixer of substances, particularly the organic. It is the carrier of upbuilding vital forces, looked at materially, though not in the same sense as oxygen, which presses into living manifestation out of macrocosmic realms of spirit. Sulphur is rather a uniting force that prompts cosmic essences to work together in building up matter. It gives itself over wholly to organic life, promotes its physical functioning, and in this way keeps it clear of infringements from the side of consciousness.

Sulphur thus plays a very important role in metabolic processes. Everyone knows how harmful emotions such as anger, fear and worry can be at mealtime. Digestive upsets are sure to result. Sulphur in the appropriate dosage is a common remedy for such afflictions, for it supports digestion by guiding nutritional elements, especially protein, over into the life of the organism in the proper way. It is active wherever a too close connection of the emotions and consciousness with the vegetative-physical indicates a need for a releasing agent; in cases, that is, where we need to have more life in our physical processes, with no interference from the activities of the soul.

The fact that sulphur has this effect is shown when an excess of it causes dizziness or a dimming of consciousness. This same effect makes sulphur useful in treating insomnia. For sleep is an extreme bodily condition in which the whole organism behaves as the metabolic system normally does. Soul and consciousness are driven out and separated completely from the purely vegetative-physical life-processes, so that these are left free to do their up-building work undisturbed.

This gives sulphur a dual nature such as the centaur has, except that it is the reverse of his. The centaur's higher self saves him from becoming wholly hardened in animality and lifts him towards the light. He longs to cast off, to reject, his lower nature. The dual-natured sulphur, however, inclines to the purely vege-tative, rejecting the higher qualities which lead towards the development of consciousness.

When we walk through a blossoming meadow in June, the month when the sun is in the constellation Gemini, we can feel the sulphuric element rampant in all the sprouting and flowering of nature. The soul of nature slumbers like the Sleeping Beauty, in the midst of all this vegetative burgeoning. In their blossoms, plants come into touch with the soul-sphere which is the source of consciousness in animals, but with the aid of the sulphur activity within them they keep the soul-sphere from penetrating more deeply into their organisms. Otherwise, as we know, they. would become poisonous.

When, in December, the sun passes through Sagittarius, almost every physical trace of plant life has disappeared; tiny, mineral-ized seeds are all that remain. But the being of the plant has withdrawn to live in the luminous heights of the realm of archetypes, where it reaches a culmination at the time of the winter solstice.

Thus we see how the processes active in sulphur and magnesia lay hold of and penetrate man and nature, having originated respectively in the constellations of Gemini and Sagittarius.

CHAPTER TWENTY-FOUR

The Oceanic Cross

We described the halogens and alkalis, magnesia and sulphur in such a way as to show that their qualities are polar opposites, yet complementary. The macrocosmic processes in which these substances originate may be traced to four regions of the cosmos: Pisces-Virgo and Sagittarius-Gemini. They form a cosmic cross.

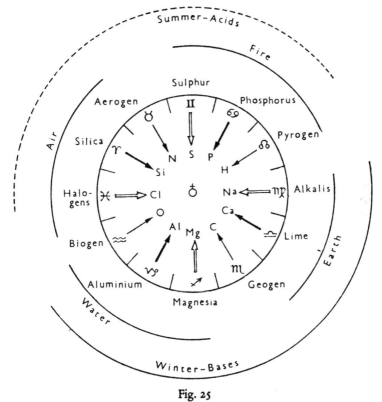

Fig. 25

The Hydro-sphere-cross
(Alongside the Mineral cross and the Atmospheric cross).

Just as we could speak of an atmospheric and of a mineral cross in the Zodiac, we can call the cross made by Pisces-Virgo and

Sagittarius-Gemini an oceanic cross. The substances that make up the atmospheric cross and work from the air mantle of the earth create the forms of organic nature; the substances of the mineral cross go to the formation of the earth's hard core; the substances of the oceanic cross fill the oceans with their salts.

The substances comprising the atmospheric cross, hydrogen, nitrogen, oxygen and carbon, and those that make up the mineral cross, phosphorus, silica, aluminium and lime, were correlated with the four Aristotelian elements: fire, air, water and earth. The same fourfold principle is to be found in the substances of the oceanic cross.

Sulphur's fire-nature is obvious. The halogen processes were described as making for decisive actions and as having an affinity to light, shown even by the halogen substances. These substances, too, are either gaseous or volatile. Magnesia, however, has an exceedingly strong affinity to water, despite its association with light and hardness. Its salts liquefy on exposure to the air; they dissolve easily and in tremendous amounts in water; and they are far more readily soluble than common salt. And the alkalis, representatives of the bases, belong clearly to the earth-element. This can be summarized as follows:

Aristotle	Atmosphere	Hydrosphere	Geosphere
Fire	Hydrogen (H)	Sulphur (S)	Phosphorus (P)
Air	Nitrogen (N)	Halogens (F, Cl, etc.)	Silica (Si)
Water	Oxygen (O)	Magnesia (Mg)	Aluminium (Al)
Earth	Carbon (C)	Alkalis (K, Na, etc.)	Lime (Ca)
Cosmos	Organic nature	Salts of the earth	Mineral earth

One peculiarity of the oceanic cross deserves mention. In the present epoch the sun rises at the spring equinox in Pisces. At this significant moment, sunrise, the constellation Virgo is in the west. The east-west axis thus coincides with the cosmic axis Pisces-Virgo, while the north-south axis coincides with that joining Sagittarius-Gemini. One can hardly imagine such a significant spatial configuration passing without leaving some specific imprint on the times. Even though the astronomical pattern changes in due course, it would seem that a particular configuration must leave some permanent spiritual mark on the cosmos. This can help us to understand the difference between zodiacal signs and constellations. We know that they no longer correspond, as they must have done at the beginning of our era, in the time of Christ.

We have now completed our description of all the zodiacal constellations as points of origin of those processes which come to earthly fixity and culmination in the various substances. Such substances as have not yet put in an appearance come from other cosmic spheres. We will devote the following pages to exploring them.

One further aspect of the Zodiac remains to be considered. The sun, rising in spring in Pisces, journeys through the constellations Aries, Taurus, Gemini, Cancer and Leo as the spring and summer months go by. Apart from their individual characteristics, these constellations have the common feature of corresponding to acid-forming substances: the halogens, silica, nitrogen, sulphur, phosphorus and hydrogen (in the form of the H-ion) are all acid-formers. Only in autumn, when the sun enters Virgo, does it begin transmitting base-forming forces, with the alkalis. This goes on until the end of winter by way of the base-forming lime, carbon (as organic base), magnesia, aluminium, and oxygen (as the OH-ion). Where the two worlds of bases and acids meet, we find intermediate stages, such as the more or less neutral hydrogen (at the end of summer) and oxygen (at the transition of winter over into spring). And the amphoteric nature of aluminium is characteristically found between the bases and the acids, between the summer and the winter halves of the Zodiac.

CHAPTER TWENTY-FIVE
The Metals

If one starts from Goethe's way of looking at nature as a whole and makes it the basis for new studies, pursued with an open mind and some artistic sensitivity, one finds that many phenomena will appear in a new light and will also reveal significant connections with one another. In this spirit we will go on to examine the special characters of the various metals and their relationship to other earthly substances.

If we come across a piece of quartz, calcite, marble, or some other crystal, we usually find it composed of the same material or chemical elements as the region in which it occurs; it is just a particularly pure, finished form of the native rock. But ores are quite another thing. These, with their metal content, run through the earth in narrow veins. They cannot therefore be looked upon as building materials like lime, silica, aluminium and phosphorus. The relation of metals to the earth and to man is completely different from that of non-metallic substances.

Metals appeal to us because of the warm, responsive, lively qualities they exhibit in their resonance, their colourful glittering, their conductivity for warmth and electricity. Stone and crystals, indeed the whole category of minerals, are by contrast insensitive, silent, immobile. True, the non-metallic minerals do achieve crystalline transparency; they can be clear and noble, as only a quartz crystal can be. But they keep to themselves and enter into no intimate relationship with anything. Their pure forms are perfect and remote.

It is the mobility, the inner liveliness of metals, that attracts us to them. When we happen on a piece of pyrite or some other metal, it affects us quite differently than do the non-metallic minerals. The former touch and move us with the soul-like fire in their sparkling and their ringing tone, while the latter lead us to marvel at their mathematical forms which remain unaltered for thousands of years. Wherever metals are, there is found the wonderful unrest of work and effort.

That is why metals have always played such an important part in the history of civilization. Various epochs have even taken their names from the metals then in use. Metals provide human tools; they can be forged and cast, hammered and drawn. This, together with their resonance and their conductivity, has been of immense significance in the cultural development of humanity. The earthly minerals seem stolid beside them. They have neither resonance nor conductivity, nor are they malleable.

The properties of the metals are more closely connected with our own nature than at first appears. They are deeply related to us and to our progress through time. Indeed, one could say that the non-metals stand to the metals as inarticulate objects do to the feeling and music inherent in the human soul.

This same contrast appears again in the basic elements of human speech. Creative forces are reflected in the fundamental structure of language. In earlier times people felt this and knew how to value the deeper significance of the spoken word. Nowadays language is simply a means of communication, but this was not so in the past. A dying man's curse or blessing held real power, for in the spoken word lived the same creative forces that built and maintain the cosmos and gave rise to our earthly substances as their final precipitation. Today, we have scarcely an inkling of the real power of the word. But those who have some feeling for what language does can still recognize the power of spoken words in contrast to a written communication.

Language has an element that we can feel to be structural or formative: the consonants. Vowels confer a different element: movement. They give 'voice' to language, the varying flow of sound that maintains its tonal continuity. Through vowels we express our innermost feelings. We use pure vowel-sounds when we exclaim 'Ah!' or 'Oh!' or 'Oo!' But when we want to describe what goes on outside us in nature: reverberating thunder, the crackle of a wood fire, the crunching of ice, the sound of a stone plopping into the water, we use consonants, for these are related to nature's elements, to voiceless objects. Consonants are crystallized, rigidified expressions of an impersonal world of form beyond the human; vowels are carriers of mobile personal feeling. One might say that while consonants form the body or the bony framework of a word, vowels are its soul and heart's blood.

Minerals and metals bring these same characteristics to expression in another way. The relationship of metals to minerals is like that of vowels to consonants.

In terms of their characteristic properties the most familiar metals can be ranged, surprisingly enough, in a regular sequence. So we find that silver has the best conductivity, the purest tone and the finest lustre; copper and mercury follow in that order. Silver's place at the head of the list is due to qualities that account for its wide use as a coating on mirrors and in strings for musical instruments; it is known to produce the purest, clearest tones. The metal least endowed with these properties is lead, closely followed by tin and iron. One would scarcely think of associating tone or lustre with lead, and it is such a poor conductor that it melts before any appreciable amount of warmth or electricity can pass through it. This makes it the best material for fuses in electrical circuits and heating installations.

Gold holds the balance between these two groups of metals. The following table shows the comparative degree to which each metal possesses resonance, lustre and conductivity for heat and electricity:

	Resonance	Lustre	Conductivity	
			Heat	Electricity
Lead				
Tin				
Iron				
Gold				
Mercury				
Copper				
Silver				

A study of the relative malleability of these metals shows them in the same order, silver and copper being the most malleable. But if we try to cast these pure metals, we shall fail. The trouble is that they absorb several times their own volume of air when molten, only to let it escape as they harden. Cast silver and copper are full of bubbles and holes, making their surfaces resemble craters. Foundrymen say that silver 'splutters'.

Lead and tin, however, lend themselves to casting, as many readers who have taken part in fortune-telling games involving lead-pouring will know from their own experience. But both resist hammering, becoming brittle and foliated.

Here too, gold occupies the middle position, for it can both be cast and hammered. The works of goldsmiths of antiquity and medieval times bear out this fact, as does the remarkable skill with which gold casts are used in modern dentistry. Gold would not make good fillings if it 'spluttered' as silver does. Iron and mercury are close to gold in lending themselves to both casting and forging. We are as familiar with wrought iron as with cast iron. Mercury's malleability is not as well known, since most people have seen it in a fluid state only, but it takes the shape of a mould on freezing and hence can be cast. And these frozen casts can also be hammered. Thus we arrive at the following table:

$$\left.\begin{array}{l}\text{Lead}\\\text{Tin}\end{array}\right\}\text{can be cast but not forged}$$

$$\left.\begin{array}{l}\text{Iron}\\\text{Gold}\\\text{Mercury}\end{array}\right\}\text{can be cast and forged}$$

$$\left.\begin{array}{l}\text{Copper}\\\text{Silver}\end{array}\right\}\text{can be forged but not cast}$$

As we see, the dynamic properties of metals give the same sequence.

This noteworthy fact is confirmed again in considering cosmic aspects. After our exploration of the earth's relationship to the starry heavens and the constellations, this will not seem strange. What, then, is the cosmic origin of the metallic processes?

We find in the planetary spheres a sequence similar to the one discovered in relation to the properties of metals (Fig. 26).

We have already explained that there is no conflict between the Ptolemaic and Copernican systems if we consider planetary paths from the planetary-sphere standpoint.

Every planet travels the heavens at its own particular speed, which can be expressed mathematically in terms of its angular velocity. If we look at Venus through a telescope some evening at six o'clock, and then try to find it at the same spot at six on

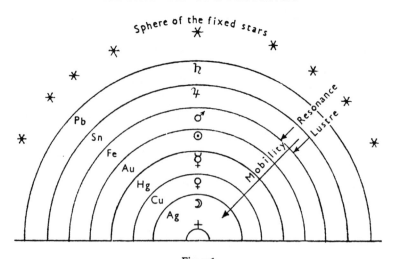

Fig. 26

Spheres of the planets against the background of the fixed stars.

the following evening, we shall go astray. Venus has moved ahead, and the telescope has to be adjusted accordingly. The angle thus described is the daily angular velocity. The average angular velocity was calculated from astronomical data for the years 1916–1934 and expressed in degrees of movement for every thirty-day period. The resulting figures are surprisingly correlated with those found in physics textbooks on the conductivity of metals, as appears from arranging the latter in the order now grown familiar and juxtaposing them with the lists of planets.

	Conductivity for Warmth + Electricity		Angular velocity of the planets in degrees	
Lead	8	10	Saturn	2
Tin	15	13	Jupiter	4
Iron	17	20	Mars	18
Gold	53	73	Sun	30
Mercury	(68)2	(76)2	Mercury	36
Copper	74	77	Venus	32
Silver	100	100	Moon	392

We see, then, that planetary movement is metamorphosed into the properties of earthly metals. The impetus of the planets appears in a metamorphosed way as conductivity. The planets move on their rounds in great curves and loops against the immobile background of the fixed stars. Just so do veins of metal run through the body of the earth, and vowels sound through the consonant structure of words and syllables.

Metals are thus wholly different from other minerals. And the difference is qualitative, like that which distinguishes planets from fixed stars, vowels from consonants.

The reason why quicksilver does not quite fit into the table above will be made clear later. The figures shown in parentheses indicate the conductivity of solid mercury.

There is a further fact that throws light on the special nature of metals as compared with all other substances. The usual arrangement of substances in the periodic system leaves no real place for metals. Some of them are put into the eighth group, where they have an isolated existence unrelated to anything else. The rest interrupt the series of related elements, and are therefore set off by themselves in side-columns. Copper, silver and gold interrupt the series of alkalis in the first group, quicksilver the alkalis of the second group, and so on. These facts will be gone into more thoroughly in a later discussion of metals of the iron group (Chapter Thirty-three). But it must be obvious already that metals possess qualities that differ radically from those of earths and other non-metallic minerals.

CHAPTER TWENTY-SIX

Copper

The name 'copper' (cuprum) comes from Cyprus, the Mediterranean island sacred to Venus. The Greeks obtained their copper from there.

A visitor to a natural history museum interested in finding the mineral with the most glorious display of colour will certainly end up in the copper section, for the various ores have a shining splendour that ranges from blue and green to red and purple. Malachite, azurite, dioptas, chalcopyrite and bornite vie with each other in the glory of their colours. Few substances possess such beauty. We can understand ascribing this metal to Venus when we see what heavenly beauty is brought here into earthly form.

Copper ores – especially the blue and green types – show by their colour that they have a special affinity to water. And most copper ores contain some water: malachite nearly 9%, azurite 6%, asperolith (a copper silicate) as much as 29%. The soluble copper salts, which form such marvellous blue and green crystals – copper sulphate (blue vitriol) or copper chloride, for example – contain up to 35% water. When heating removes the water, the colour and the crystalline form both disappear. Malachite breaks up into a black powder, while blue vitriol disintegrates into a white powder on contact with dry air. This powder has a great attraction for moisture, and on absorbing it quickly returns to its familiar dark-blue crystallized state.

We may say, then, that copper absorbs water and changes it into form and colour. This capacity for organizing fluids is a characteristic copper attribute. Plants perform this function in organic nature. Plant matter, which is 70% to 90% water, is permeated by formative forces which organize it into form and colour. Plant fibres are really condensed fluids. Copper, which has the same tendency in the dead mineral world, is closely bound up with all vegetative processes. We know that treating cut flowers with copper keeps them fresh longer. This is often done by putting copper coins in the vase with them. Certain copper ores even imitate plant forms. Malachite, for example, tends to veined, leaflike surface patterns, and pure copper in its natural state looks like tiny trees or leaves.

Chemically, copper is as versatile and many-sided as it is lively in variety of forms. We see this in the great variety of copper ores. Since the metal is so readily soluble in almost any acid, there are very numerous, beautifully crystallized and gloriously coloured salts. Green and blue predominate. Copper is chemically so active that it combines with most other substances. Avogadro's law states that individual elements combine only in simple and multiple proportions, resulting in simple molecular formulae. But copper often seems to contradict this law by forming so-called complex salts. It combines, for example, with ammonia, with the alkaline salts of organic acids, and with sugar and other carbohydrates to form groups of substances that still present unsolved riddles. Despite painstaking research and much theorizing, of which Werner's theory of partial valencies is an example, the molecular structure of such complexes remains more or less obscure.

An ammoniacal copper solution – itself a complex compound – dissolves cottonwool and even wood, thereby creating a still more complex compound. When this is forced through very fine tubing into a dilute solution of sulphuric acid, the acid destroys the complex, causing the threads of liquid squirted into it to solidify into cellulose again. One of the artificial silks (Bemberg silk) is made by this method.

We see, then, that copper cares little about the logic of chemical laws. Its activity belongs to the unpredictable realm of life, where change and surprises are the order of the day. Such is the nature of the Venus element wherever it is found. The planet itself exhibits these characteristics in the path it travels, in the way it alternates its roles of morning and evening star, and in its changing phases.

Copper is thus the substance in which is manifest the cosmic process responsible for stimulating and maintaining a lively circulation of the liquid element. Hence it supports the various vegetative functions of the human organism.

Chief among these are digestion and blood circulation. To

avoid misunderstanding, it must again be emphasized that when we speak here of a copper activity, we mean a higher than material process. A therapeutic use of copper therefore confines itself to the higher potencies which are sufficiently dynamic to have a curative effect in cases of fatigue or where regenerative processes need stimulating. We will not attempt any further elucidation here of the other special or wider connections of the Venus or copper forces with the human organism. We are concerned solely with showing how the properties of copper are characteristic of that process which the ancients associated with Venus. The copper process, with its enlivening and organizing effect on fluids which we witness in its creation of the surprising complex compounds; the beauty of the various material forms of copper; indeed, its typical red-gold – all these are earthly reflections of the Venus nature.

When copper is separated from the fluid element that permeates it and reduced to the pure metal, we see the pure red metallic copper appearing and can understand why alchemists said of Venus that she had a blue cloak and a red spirit. Great artists of the past either knew or sensed the truth about such matters. So we find Botticelli painting his sea-born Venus with hair of shining red and making her rise from blue-green ocean waves. Even flame repeats this dynamic colour phenomenon, for copper burns with a shining blue-green fire, and here and there a flash of red-tipped flame.

CHAPTER TWENTY-SEVEN

Tin

Tin is in many respects the opposite of copper. It occupies an extremely modest position in the family of minerals. There is only one important tin-ore: cassiterite. It is very plain-looking. It can be transparent, but is more frequently dark brown or black. It is found in regular, almost spherical crystal form, in granite or quartz. Only attentive observers with an eye for fine detail will notice the fluctuating play of colour in the depths of these dark crystals.

Tin, unlike copper, has a strong aversion to the fluid element. Cassiterite is completely dry, and tin compounds produced in the laboratory are more apt to rid themselves of water than to absorb it. Orthostannic acid, for example, tends to change into metastannic acid, which contains less water. This is accompanied by densification. The condensed particles adhere (polymerization) and are precipitated; dehydration proceeds here by way of polymerization to precipitation.

It is worth noting in this connection that tin is found almost exclusively in islands. The Phoenicians got their tin in Cornwall. Nowadays most of our tin comes from islands of the Malay archipelago. There are tin deposits in other islands or peninsulas, Japan and Tasmania. Peru, the only important source of tin not situated on a peninsula or island, is a high country that geographers and geologists think may well have been an island at one time. World tin production in the past thirty years averaged one hundred and twenty thousand tons annually, and eighty thousand tons of it came from Malaya.

Tin, which unlike copper withdraws from water and is characteristically an island metal, seems to harbour a cosmic force that conjures plastic forms out of fluids. Its dynamic is thus the very opposite of copper's. Copper is at home in the living stream of circulating juices; like plants, it reaches for the light and brings forth colour through a refining, etherealizing process that is all upward movement. Tin, however, dislikes water, condenses fluids into solid forms and has a drying action on them.

Its process is one of descent, its action like that of warm air in drying up moisture and bringing out the element of form.

A further significant phenomenon is associated with these processes of condensation and coagulation. The fluid extracted from blossoms – blue flax, for example – is pale and well nigh colourless. The addition of a drop of tin-salt solution transforms it into the most glorious purple. That is why tin has always been indispensable as a mordant in dyes and is still important in wool and silk dyeing. We see that tin, itself colourless, has the effect of bringing out latent colour. If we take the Goethean approach and view colour as a harmonious interplay of light and darkness, we shall realize that tin has the organizing power to relate the two.

The same forces are at work in the synthesization of the 'purple of Cassius'. Essentially this is colloidal gold, but tin was required to get the deep purple used in medieval times for staining glass.

Copper, then, belongs with the fluid element and with chemical action, which reaches a high point in the glorious colours of its salts. Tin is primarily a light and air organizing force; it works down from these realms into the fluid element, creating form there. The chemistry of tin, in contrast to copper, is as simple and logical as can be; it contains no surprises.

It is not hard to find signs of the tin process in the human organism. We may look upon it as the sculptor who works from above downward, from the finer towards the denser. While the copper process works enliveningly on circulation, etherealizes nutrient substances in the final stages of digestion and helps to merge them with the respiratory process, tin is active at the opposite pole in helping the forms of the various organs to coagulate out of the body's colloidal fluids and condensing these into the cartilaginous substance of the embryonic skeleton. Tin builds dams to keep water in its place, as it does for example in the formation of the brain. Hydrocephalus is an illness caused by tin deficiency, and is often accompanied by a too soft condition of the bones. In speaking of tin we do not, of course, mean the substance, but its formative forces. The above facts clearly indicate the therapeutic use that can be made of high potencies of tin.

Tin is also used as a solder. What is the soldering process but a joining of two pieces of metal with the help of tin? This points directly to a further function of tin forces in the human organism. Tin links bone with bone through the agency of the cartilage in our joints, and on a higher level is active in the capacity of the mind to link thought with thought in logical sequences.

Tin conjures forth colour in colourless plant extracts, inducing coagulation in the substance. The brain is the physical counterpart of the light-filled world of thought. Like the eye, which is formed by light for the purpose of perceiving light, the brain is built by thinking for the purpose of perceiving thought, and the Jupiter-tin process is its agent.

Looking back into antiquity, we find a wonderful picturing of this process in Greek mythology. Zeus-Jupiter is shown enthroned in the clouds, ruling over light and air. His throne of clouds is simply coagulated water-vapour. Now mythological beings are not simply personifications of nature-forces, nor is the concept of divinity merely a glorifying of the nature-forces before they were physically understood. Such a view of the matter follows naturally from a materialistic world-conception. But the truth is that the humanity of earlier times was guided by a divine world. Only at about the time of Christ did this change, for then these divine powers began to withdraw from man's surroundings and to enter the human soul and work within it. What had been divine wisdom became the power of human thinking. This thinking still lacks the capacity to see things in their wholeness; it is still enmeshed in the chaos of the partial, physical approach. The throne of Jupiter is no longer in the world of wafting clouds, but in man's brain, where the view is obscured by physical objects. And when our thoughts are not warmed by the fires of enthusiasm they freeze to 'grey matter' and fall apart in atomistic abstractions.

This is the fate of tin, too, on long exposure to freezing temperatures; it disintegrates. This seems like an illness, even an infectious illness, for if one scratches a piece of 'healthy' white tin and strews 'sick' tin dust over it, a pustule appears and exudes a grey powder. This phenomenon has been appropriately termed 'tin pestilence'.

The tin-Jupiter process is thus the opposite pole of the copper-Venus process.

CHAPTER TWENTY-EIGHT

Lead

Lead, which comes first in our list of metals, has the least lustre, a dull resonance and almost no conductivity. Outwardly it is even plainer than tin. It is heavy and dark grey; one might almost call it gloomy, with a moribund gleam. It is completely moistureless, with an even more negative relationship to water than tin has. Though soft, it is brittle and therefore not malleable. Lead ore lacks the slightest moisture content, and the soluble salts of lead crystallize with no water of crystallization. As we have seen that water is the basis of all life, we can see that lead's heavy, gloomy aspect has a relationship to death.

But if we pick up a piece of lead, we are surprised to find it feeling softer and warmer than one would have expected of a metal. It even feels strangely oily. For all its plainness, lead apparently possesses unsuspected qualities. And if one goes on to make a closer study of it, one comes to know another, most important side of lead which has nothing to do with heaviness: the fire that lives hidden in its depths. Some lead ores show it quite externally.

Though the chief lead ore, galenite, has lead's typically gloomy look, there is a whole series of lead ores whose bright colouring betrays the fire within. Yellow, orange and red occur most frequently. Croconite and wulfenite, the red and yellow ores, and several others, sparkle as though fire itself had fashioned them. White lead ore, cerussite, though a colourless white, brings hidden fire to expression in the way it is shaped. It is built of sheaves of needles, or is a network of glittering laminae. It looks amazingly like bone structure. Thus lead unites two very strongly contrasting forces: rigid heaviness and revivifying inner fire.

The fire-nature of lead is beautifully illustrated in an experiment which we will describe in some detail because it is generally so little known. If we want to pulverize lead to a very fine powder, we have to work in a vacuum, for the powder would otherwise ignite and gradually burn down to bright yellow ash.

Although this pulverizing is impossible on account of the

sticky consistency of lead, there is another way of achieving the same end and producing pyrophorus lead. The technique is as follows. Lead citrate is put in a glass tube sealed at one end and gently shaken till it settles at the bottom. The open end of the tube is attached to a vacuum pump. The citrate is slowly broken down with the heat of a small flame, while the vapour and carbon monoxide thus generated are drawn off by the vacuum. The end product is metallic lead in the form of a fine powder.

When the tube is sealed with a glass cock, taken off the flame and the pump disconnected, the lead can be kept for weeks or months, provided the seal is tight enough. But the moment air gets in, the lead bursts into flame and is gradually consumed.

The chemistry of lead, like that of tin, is very simple and straightforward. Its lack of affinity to the lively chemical activity of water renders it chemically sluggish. This quality makes it excellent pipe material.

Bells made of bronze with a lead alloy ring with a warm depth of tone. But impermeability to rays of energy such as those given off by X-rays and radium is a special characteristic of lead and makes it particularly well-suited to serve as a shield against their destructive action.

Lead's double nature, its dead weight counterposed to living fire, together with its shielding properties, can easily be understood when we consider its cosmic source of origin: Saturn.

This planet has two characteristics which are at once apparent to telescopic observation – its dark core and the bright ring that encircles it. And its distant orbit encloses the whole planetary system, sheltering it from cosmic radiation. We forget all too readily that the life of earth and its creatures is made possible and maintained in a carefully attuned balance of forces by the sheaths that surround it. Earth has its hydrosphere, its air and warmth mantles, and its ionized layers, and beyond these the planetary spheres. The last and most important enclosure, which separates the planetary system from the rest of the cosmos and makes it an independent entity, is the Saturn sheath; or, we might say, the lead process. And when forces such as X-rays and radium appear in the earth sphere, it is lead again that protects us from these deadly energies and enables us to live independently of them.

Lead's protective function appears very clearly in the smelting process. Lead ores are always found in conjunction with silver, and the smelting process produces a conglomerate with a good deal of silver in it. This is then heated, and the more volatile lead goes off in smoke. The percentage of silver in the residue keeps on rising, until finally there is only a mass of liquid silver mantled in a film of lead, which protects it from contact with the air and thus prevents spattering. When, finally, the lead skin becomes too thin and tears (the silver 'peeks out', as the smelter says), the lead has done its job and the last of it goes up in smoke.

If we look for the Saturn-lead process in the microcosm, man, we come upon the same comprehensive functions. One of lead's activities is building bone, which involves the death-process of mineralization. The densification brought about by tin in the organism is carried only to the stage of cartilage, which is still plastic. It is lead that carries mineralization to its real conclusion, which one might call a kind of death. But this death enables man to be at home in the realm of gravity, as well as to assert himself against it.

Bone-building is the final stage of the lead process in the human organism. Man incorporates death into himself with his bony structure. But at the heart of this mineralized precipitate of our physical selves, at the core of what is most dead in us, we find the scene of life's creation: the red marrow, where new blood, new red blood corpuscles are made.

The lead process is thus linked with processes of death and resurrection in the organism. On the one hand it lends itself to the forces of mineralization active in shaping our bony frame-work, while on the other it supports independent consciousness, enabling us to co-ordinate our perceptions and relate them to the ego, the centre of each man's personal universe. The lead and silica processes are similar in their influence on this functioning of the senses, which is based on a constant breaking down or dying. Disintegration is continually going on in us, particularly in our nervous systems, but it is just because we are always ex-periencing partial death that we can become conscious beings. At the moment of complete and final death an unimaginable enhancement of consciousness occurs owing to the sudden setting free of so many formative forces from their bodily tasks. Reports to this effect have been made by individuals who came close to death but were revived. They speak of seeing a tremendous panorama of their whole life, such as cannot be experienced normally.

The lead process is thus related to the most spiritual as well as to the most material aspects of our being: to that maturity of consciousness which manifests itself in the warmth of an all-embracing human understanding. When the lead process within us is thrown out of balance, we lose the firm footing of an ego-directed soul life, as can be witnessed in the poor memory of sclerotics or in the brittleness of bone and the failing senses of old age.

The wide range of properties inherent in lead makes it a valuable medicament in cases where, on the one hand, the process of densification proceeds abnormally, or on the other, consciousness is disturbed in the ways just described. Used in high potencies, it has given good results in the treatment of various sclerotic and related conditions.

In ancient times, this twofold character of the Saturn-influence was well known. Saturn was regarded as the representative of death, of the forces responsible for aging, but equally of the deepest attainable wisdom. The Greeks looked upon Chronos-Saturn as the creator of time. But time harbours new beginnings as well as endings. Time is known as the great healer, the overcomer, who offers resurrection after every death.

It was natural to fear Saturn, for this border-guard who constantly patrols the boundaries of our planetary system is also the guardian of a treasure of ultimate knowledge, to be attained only at cost of the greatest sacrifice. Suffering and loneliness are awakeners of knowledge.

Lead is in a profound sense the final stage of a great cosmic evolution.

It is clear that this way of looking at the world leads to a new relationship to the substances of the earth that does justice to the spirit variously active in them. We are not concerned with merely reaffirming insights of an earlier time, but rather with building new bridges to an understanding of the nature of substances. We can take literally Goethe's mature saying: 'All that is fleeting is merely a symbol.'

Even the physical planet Saturn is a powerful symbol of the lead-Saturn process, with its characteristic twofold creativity, for Saturn is made up of a core and an encircling ring. On every level, Saturn builds both core and enclosure. It provides both physical and spiritual bases for the ego and its 'self-contained' world of individual consciousness.

In the bone-building process we come up against the mineral world inside ourselves. In sense-perception we confront matter in the outside world. Both are boundary-experiences and are guarded by Saturn. Here too he maintains his shielding character. Present-day consciousness is precluded from looking behind mineral matter: we come up against its resistance.

In our inner life, also, we are shielded by our senses from becoming aware of the deeper layers of the soul, which we are normally no better equipped to deal with than we are with the energies set free by the disintegration of matter.

A striking phenomenon in this connection is that lead is the only substance able to hold out unchanged against radioactivity. Thus it proves itself to be the most stable and solidly resistant of all earthly substances.

CHAPTER TWENTY-NINE

Iron

As we have already shown, metals play virtually no part in building up the earth, since they are merely threaded through the body of the earth in veins and give it a certain life. This holds true of iron, too, whether it occurs in larger compact masses (e.g. at Erzberg in Styria, where a whole mountain of it is gradually being reduced in open-face mining operations) or in the small quantities found scattered all over the earth. If minerals such as lime, granite or clay have a brownish tinge, this comes from an admixture of tiny iron particles. There is scarcely a spring, a river or a lake without a detectable amount of iron in it. Its universal presence in the earth reminds us that it also permeates the human bloodstream. It is the only metal present there to an appreciable degree, and is one of the most important in the body, although it cannot be called a true body-building material.

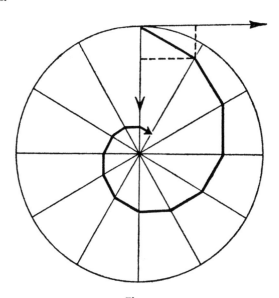

Fig. 27

The spiral as a resultant of spherical and radial forces.

The universality of iron makes for great variety in its ores. A morphological study reveals a notable structural variety, with regard not only to the shapes of single crystals, but to the overall crystalline arrangement or patterning. Two dynamic tendencies stand out. On the one hand there is a distinctly radial structure, apparent especially in marcasite. On the other hand there is a tendency to the tangential in certain spherical formations such as hematite, limonite, and the like. Spirals result from an interplay of these two dynamics.

One would expect to find iron ores in which spherical and radial tendencies have combined to make a spiral. And we do find them. Wonderful spiral forms are to be seen in many samples of siderite, and the so-called iron rose (*Eisenrose*) owes its spiral pattern to the same combination of forces.

Some may wonder whether the radial structure of marcasite is the product of centripetal or centrifugal forces. But when we observe its spherical surface and note that it often has a hollow centre or is filled with some other kind of stone, such as lime, we realize that only a radial force working from a periphery inward could have shaped it thus. Considering, moreover, that meteors (and marcasite must be of similar origin) are to be thought of as cumulative accretions, it becomes still clearer that this ore could have come into being only through a centripetal dynamic.

If we compare marcasite with antimonite, which also has a radial structure, we can clearly see the contrast between the centripetal and centrifugal forces working in them. The antimonite crystal radiates out from a centre, and is easily recognized as the product of centrifugal forces.

The spiral tendency always arises when time enters space and develops towards a centre. The fact that this dynamic shows up so clearly in iron ores points to the fundamental role played by the iron process, for it transforms spherical forces quite unrelated to the laws of earth into radial forces working towards a centre. Or we can say that the function of iron is to help cosmic, weightless elements to enter the sphere of gravity. This is a characteristic of iron to be found at every level of its functioning.

The same dynamic operates in the strange phenomenon of the

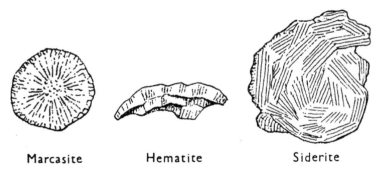

| Marcasite | Hematite | Siderite |

Fig. 28

Radial, spherical and spiral form-tendencies of iron-ores.

formation of meteors and their entrance into the atmosphere. The cosmic iron process of the Mars sphere can be seen at work here, right down to physical manifestation. Iron is the only substance which makes visible in an archetypal picture the incarnating force during its spiral descent, for the path followed by meteors is indeed a spiral, the result of interacting radial and spherical forces.

The following phenomena serve as further illustrations of the functioning of iron as a carrier of incarnation forces. Iron has the closest connection with magnetism. Iron filings strewn in a magnetic field quickly arrange themselves along the lines of the

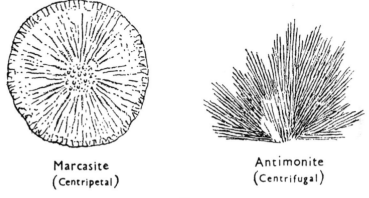

| Marcasite
(Centripetal) | Antimonite
(Centrifugal) |

Fig. 29

Formative forces radiating inwards and outwards.

flow of force between the poles. This is a phenomenon reminiscent of the Chladnian tone figures mentioned above in connection with the ordering force inherent in music. It almost seems as though a like force resides in both music and magnetism. But we have to remember that magnetism works out of subterranean depths of earth, and that iron is almost the only substance – together with the condensing force of electricity – that reacts strongly to these sub-earthly forces.

Iron, then, conforms to the force-patterns of the earth's magnetism. This conformity comes to good use in the compass needle that helps us to orientate ourselves in space. But we need inner orientation, too, and inner adjustment to spatial forces. It is the iron in our blood that enables us to anchor our personalities in our bodily processes. Without iron we would quite literally lack 'presence of mind'. Our human egos give our lives their orientation; the iron in our blood serves as a mediator, relating the ego to the spatial dynamics of the earth. It provides a basis for our earthly activity and creativity. Do we not say of a determined person that he 'has steel in his blood'?

These characteristics have corresponding applications in therapy. Iron in high potencies is a good remedy for persons who have trouble in mastering their bodily processes. We see this difficulty in cases where processes of a certain kind drive the personality out of the body or cause a diminution of consciousness, as happens most signally when the sulphur forces of digestion gain the upper hand. We have often called attention to the fact that consciousness can develop only by depressing biological activity; that it rests upon a death process. The torpid state induced by the sulphurous nature of the digestive process is continually being lightened up by iron as it beats down the sulphur process.

We see a corresponding process in the mineral world in iron's readiness to combine with forces of the earth's depths and with the fire-related sulphur to be found there. It is to this we owe the harmonious pentagon-dodecahedron form of iron pyrites, the most perfect of all crystal shapes.

Iron is also virtually the only metal with a close relationship to carbon, 'earth substance'. Carbon is not only the basic building stuff of all organic matter but plays a role in the mineral kingdom too, as for example in lime (calcium carbonate). It is significant in this connection that it is precisely iron which has such a close bond with carbon, combining with it, dissolving or otherwise transforming it.

Everyone knows that our machine-age civilization owes its very existence to this relationship of iron to carbon. Pure iron is soft and malleable, and therefore unsuitable for use in machine parts. But in a molten state it greedily absorbs carbon, which dissolves in it as readily as salt in water. The product, on solidifying, is no longer soft; it has become hard, brittle cast-iron. The earth-substance, carbon, that now permeates it lends iron hardness and permanence of form.

Steel is an intermediate stage between malleable iron and cast iron. It is not as hard and brittle as cast iron is, but tends, on being poured, to try to resume its original form – which means, in effect, that it is elastic.

When carbon is dissolved in white-hot molten iron and the mixture is solidified by sudden cooling, the resultant iron has tiny diamonds distributed all through it. It is highly stimulating to live through in imagination the transformation of dark carbon into these rare and shining jewels. Is it not conceivable that man's spirit might some day learn to use the forces of iron to change the earthly nature of his physical body into something far more related to the realm of light?

The iron processes in the blood are rhythmical, reminding us that life is so to speak elastic, like steel. But when the forces of embodiment work too strongly and life tries to exceed its normal span, the vital processes gradually become mummifying forces. Some of the industrial uses to which iron is put will serve as illustrations.

Iron is used for tanning leather. But is leather not mummified skin? In earlier times almost all tanning was done with oak bark. The entire habitus of oak clearly indicates that an iron process is at work here. This gnarled and stubborn tree, with its hard wood and the thick, tough bark that is a perfect picture of mummification, certainly expresses the nature of iron. Trees, too, reflect the great fundamental processes that weave the tapestry of nature before they come wholly to rest in substances. And we must emphasize again that it is not the substance iron we refer to here, but rather the dynamic properties of iron active in the tree's life-processes.

A few decades ago, tanners switched from oak bark to chromium salts, especially in tanning vamp leather. More recently still, iron salts themselves came into use. We will go later into the reasons for this use of chromium; it is a 'brother' of iron and very like it in several respects.

All this indicates that iron becomes a mummifying agent when it overshoots the mark. We have already pointed out that the substance iron is almost universally distributed. The same universality appears in its chemistry.

Apart from those qualities which have been described as relating iron to forces purely of the earth (magnetism and carbon), when we examine it from the chemist's standpoint we find it possessing traces of other properties noted in connection with our discussions of copper, lead and tin. For water, air and fire forces are reflected in iron's basically earthy nature.

In our discussion of copper we spoke of the power it has to organize the watery element into shape and colour; and of the versatility shown in its lively chemical activity. Ferrous sulphate (generally known as green vitriol) is quite similar to blue vitriol in many respects. It has beautiful green crystals that contain some water. And as in the case of blue vitriol, green vitriol loses its water content on being heated or even on exposure to dry air, crumbling to a white powder. The green crystalline form is re-established in the presence of moisture.

We also find in iron the tendency to form complex salts. Iron and cyanide in combination form a group of substances, known as red and yellow potassium prussiate, which are the source of a highly coloured material called Prussian blue. Iron cannot be detected in these compounds by ordinary chemical analysis. We might say it acts anonymously in them. Many scientists have tried to explain their complex and peculiar structure and to decipher their true nature. Werner, for example, proposed half and quarter valencies as a possible solution of the problem. But the fact remains that iron behaves like copper here, effecting all sorts of unexpected transmutations. But this is just what makes it a healing element in nature and in the human organism, for it absorbs cyanide, which is poisonous, and changes it surprisingly enough into harmless Prussian blue. As the digestive process constantly produces cyanide-like poisons, we would be subject to poisoning at every meal were it not for the activity of iron in our blood, rendering the poisons harmless.

Iron's kinship with tin shows up in the following phenomenon, where light and air processes play an important role. Tin brings out latent colour, and is therefore used as a mordant in dyeing silk and wool. Iron can be used for the same purpose, though it does not produce the bright, shining colours that tin mordants do. Iron favours, earthier, darker shades, the so-called 'fashionable' colours. This shows, nevertheless, that, like tin, it has the capacity to organize light and darkness into colour.

We can go a step further and study iron's relationship to light and air through its behaviour in different valencies. We described in what sense valencies are an expression of musical qualities of matter, being subject to the law of simple and multiple proportions. There are bivalent and trivalent forms of iron. In our terms, then, iron sings in seconds and thirds. It is fascinating to observe that iron's singing is influenced by the intensity of light. Direct sunlight changes trivalent into bivalent iron, or, in other words, makes it change suddenly from singing thirds to singing seconds. This process is applied in making photographic prints from a negative. A complex cyanide compound is used. In the presence of light it changes into a deep blue called Turnbull blue. The familiar draughtsman's blueprint is made in this way.

This change in valency, which can also be reversed under certain conditions, is a characteristic of iron. Other substances possess it to a certain but not at all comparable degree, especially in relation to oxygen. This fact is of great importance in chemical analysis. The reaction of permanganate of potash and a ferrous salt, as the salts of bivalent iron are called, consists in the absorption by iron of the oxygen present in permanganate of potash, causing the deep violet colour of the permanganate solution to disappear instantly. If the strength of the permanganate solution is known, the quantity of iron can be calculated.

This oxygen absorption, with the simultaneous disappearance of colour in the solution, is an impressive phenomenon if the experiment is done as follows. A 10% solution of iron sulphate is poured into a large glass cylinder so as to cover the bottom. To this, a deep violet permanganate solution of about 1% is added slowly, with constant stirring. The deep violet disappears on contact with the iron solution, and a huge amount of the violet permanganate solution is needed to satisfy iron's appetite for oxygen. When it is satisfied, it expels the oxygen again under certain conditions. The oxygen thus released is able to oxidize chemical compounds such as hydrogen iodide, causing the iodine thus produced to restore the colour.

$$10FeSO_4 + 2KMnO_4 + 8H_2SO_4 = 5Fe_2(SO_4)_3 + K_2SO_4 +$$

Ferrous sulphate (Second)	Permanganate	Sulphuric acid	Ferric sulphate (Third)	Potassium sulphate

$$2MnSO_4 + 8H_2O$$

Manganese sulphate

deep-red-violet → colourless

$$Fe_2(SO_4)_3 + 2HI = 2FeSO_4 + I_2 + H_2SO_4$$

Ferric sulphate (Third)	Hydrogen iodide	Ferrous sulphate (Second)	Iodine	Sulphuric acid

colourless → red (with starch blue)

This experiment gives a good picture of iron's capacity to bring air into motion. As we know, our breathing depends on this capacity of iron in our blood. Iron sings thirds in red arterial blood, which is air-saturated. It sings seconds in the blue venous blood flowing back to the heart after giving up its oxygen. On arriving in the lungs, it again takes up oxygen and changes back to thirds.

If we observe this activity of breathing in our lungs, we can hardly avoid the impression that the orbit of Mars, as it rhythmically nears the earth and then moves farther away, is a cosmic counterpart of the human breathing process.

A planet's movement in space determines its sphere of influence on the earth. This influence rays in from a circumference determined by the planet's position at a given moment. We might say that the planet is like a cell nucleus, dominating the surround-

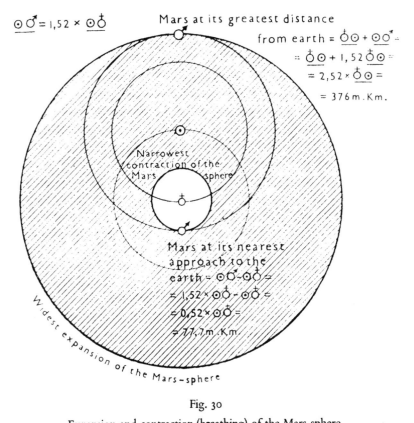

Fig. 30

Expansion and contraction (breathing) of the Mars-sphere.

ing cell area. Mars similarly dominates the space which its path circumscribes, expanding or contracting its sphere of influence exactly as though it were an elastic cell membrane. The lungs carry out the same movement as we breathe in and out, and our power of speech results from it. We showed above that the chief function of the iron process (or of the Mars sphere) is to bring us down to earth. In this context, we can picture iron as the force that conducts the musicality of the cosmos, Plato's 'harmony of the spheres', down to earth and into man's organism, where it finds its highest and most sublime expression in human speech. Speech is a power which comes from the control and articulation of sound by a sovereign and fully incarnated ego.

Finally, there is the question of iron being tinged with qualities of lead. On the one hand it controls the ossification process that brings us fully down to earth. On the other, it has lead's fiery quality, flaming up towards the spiritual worlds, for it is practically the only other metal besides lead that is capable of spontaneous combustion when ground to the point of almost molecular fineness. Many readers will be familiar with pyrophorous iron, made by reducing precipitated powdered iron hydroxide with hydrogen. Like lead, it can be kept for a considerable period if oxygen is excluded, and then bursts into flames the moment it comes in contact with the air. Is this warmth of iron not the very same force that makes our blood run faster when we are 'set on fire' by enthusiasm?

A glance at history will show the important role iron has played in man's evolution. In Roman times, during what is generally known as the Iron Age, the human ego began to enter fully into its earthly embodiment. Museums offer interesting possibilities of studying the statuary of various periods. If we compare the face of a Roman emperor with that of a Greek philosopher or an Egyptian king, we are struck by the ego-forces of individual personality which come to expression for the first time in the Roman's face. They are entirely, or almost entirely, absent from the other two faces. An observer sensitive to such things has the immediate impression that the eyes of sculptured figures of earlier civilizations look right through one to something far higher. An Egyptian gaze, though it looks right into us, does so with an expression of such infinite remoteness that we are overcome with a sense of awe, as though in the presence of the eternal. There are individuals who feel positively uncomfortable when they confront a smiling Pharaoh. But Roman emperors seem just like ourselves, with their personalities clearly imprinted on their faces.

Perhaps we should allow one exception among Greek physiognomies – that of Socrates. His expression betrays the struggle of awakening personality to the point of an almost agonized grimace. Something in him is wrestling through to birth. His countenance bears a certain resemblance to embryonic forms. How are we to understand this exceptional case? Just as Aristotle was the pioneer of logical thinking, Socrates can be regarded as an analytical sceptic who rejected divine authority when it was already becoming a mere tradition. In doing so, he laid the foundation for a later ego development. We might call him a pioneer of personality.

Roman law is the first documentary statement of human personality and of its relation to the physical environment. It marked the historic moment when ownership was put on a legal footing. Personal property was legally accorded such significance that Roman citizens could bequeath it to whomever they chose. Wills made their first appearance at this time. The birth of the principle of individual rights was now really accomplished.

Thus the Mars impulse is not to be found only in the warlike nature of the Romans and in their iron weapons (the Greeks had still used bronze); it can be seen also in the whole trend of their social life and culture. Their highly developed power of rhetoric is another example of the same tendency. The Roman orator no longer sought divine inspiration; he addressed himself to his listeners' insight and played upon their mood. This was often done so cleverly and persuasively as to override decisions of the Senate.

While the Greeks had hearkened to their oracles and heard the gods speak through them, the Romans listened to orators who depended on their own intelligence and expressed their own feelings. These impulses reached a peak with the climax of Roman civilization. Later, they collapsed, thus contributing their share of darkness to the Dark Ages. Medieval armour grew increasingly complex and hard to handle. Elaborate devices were required for hoisting knights into their saddles at tournaments, and if they were unseated while jousting, they had to lie there in the sand. This same awkwardness and decadence can be detected in language. Official style in the Middle Ages produced absolute monstrosities of phrasing.

The invention of gunpowder brought a sudden end to these

developments. Mars was dethroned. But the Mars impulse now took over technology, and Mars became the inspirer of the Machine Age.

We have seen how the Mars impulses at work in iron are carriers of the forces of embodiment. We have shown also how these forces lead to mummification if they become too active and overwhelm the vital processes. The personality can also be affected if driven to plunge too deeply into the body by the Mars forces. Individuality then becomes denatured and crass egoism results. Egoism makes irresponsible use of machines for its own ends. Whether machinery proves a curse or a blessing depends on man himself. It is a destructive force in the hands of materialists. But if man rises to a recognition of his spiritual origin, the machine will serve him and the social order as a friend.

CHAPTER THIRTY

Quicksilver

Quicksilver, the metal of Mercury, is as like iron in some respects as it is opposite in others.

It is found in nature in the same state we are familiar with from the use of mercury in thermometers: as a liquid metal. It occurs in smaller and larger globules in the matrix rock, often in the company of mercury sulphide, more familiarly known as cinnabar. The globules tend to unite and form larger ones, which shatter at a tap. Their liveliness is a wonderful sight. Quicksilver's ancient name, 'Mercurius vivus', is truly deserved. And this mobility is demonstrable in many other physical and chemical phenomena.

One of these is the narrow margin between quicksilver's freezing and boiling points. It changes quickly from a solid to the fluid or the gaseous state. It drops only 399°C. from the boiling point to freezing, as against gold's 1,537°. Mercury's quick passage through the different states makes it akin to water, which also exhibits a great variety of phenomena in the interplay of the elements, producing steam and clouds, fog, rain, hoarfrost, snow, and ice, and leading each over into the next. 'Hydrargyrum', the Latin name for quicksilver and the origin of the symbol Hg, acknowledges this kinship.

Lively, liquid quicksilver is 'young', compared with the other metals. It has retained the fluid form of earlier conditions of the earth and held out against ageing and solidifying. In the table of metals it is abnormal for being a very poor conductor (cf. Chapter Twenty-five). It is only externally lively, not having yet achieved the inner mobility of conductivity. When the temperature is lowered below the freezing point, however, its conductivity increases markedly. In this solid state it thus comes to possess all the properties which its position in the table of metals warrants.

Quicksilver's reactivity to warmth is especially notable. It expands and contracts in exact proportion to the rise and fall of surrounding temperatures. It is not a carrier of inner fire processes, as lead and iron are, for it simply reacts to changes in temperature with speed, precision and agility.

Its most significant quality is its capacity to dissolve other metals, making alloys known as amalgams. Only one metal has the power to resist this amalgamation, and that, strangely enough, is iron. For this reason, mercury is kept in iron flasks for storage and transport. We need not be surprised at iron's resistance. Mercury did not participate in the last stages of earth's densifying and remained a fluid, while the force that leads most deeply over into earthiness, transforming the cosmic into the telluric, the spherical into the radial, is that of iron.

Now, if we compare mercury with iron, searching out their likenesses and dissimilarities, we come upon a clue to mercury's globule-forming tendency: it signifies both a retaining of the cosmic sphere-form and a leaning toward individualization. Here we see a perfect illustration of the opposite directions taken by iron and mercury. Mercury's splitting up into numbers of small globules and its tremendous mobility are in the greatest possible contrast to the consolidating power of iron. Iron carries forces of embodiment which, if they go too far, lead to mummification in the sphere of life and to egoism in the soul sphere. On the other hand, quicksilver represents the force that combines small entities into larger wholes, making one large globule of many tiny ones. This is the basis of the capacity to amalgamate, which, over-developed, ends in erasing identity.

Chemistry presents phenomena that well illustrate these mercurial characteristics. The reaction of chloride of mercury with potassium iodide precipitates a glorious red mercuric iodide. An excess of potassium iodide, however, has a surprising effect: the red precipitate disappears, leaving a clear, watery solution.

$$HgCl_2 \quad + \quad 2KI \quad = \quad HgI_2 \quad + \quad 2KCl$$

| Mercuric chloride | Potassium iodide | Mercuric iodide (red precipitate) | Potassium chloride |

$$HgI_2 \quad + \quad 2KI \quad = \quad HgI_4K_2$$

| Mercuric iodide | Potassium iodide | Potassium mercuric iodide (clear watery solution) |

Chemists call this strange association of elements 'double-salt-forming'. But this explains nothing and does not even describe the nature of the occurrence, especially since the quicksilver here can scarcely be detected by ordinary analytical methods. Moreover, it takes up other substances, such as ammonia (Hg_2ONH_2I). The phenomenon can be explained only by bringing the whole nature of mercury into account. Its most characteristic chemical trait is that of association, a feature already noted in our discussion of its readiness to form alloys, or amalgams.

Quicksilver draws into chemical association substances that otherwise show little affinity for one another. Sometimes there is such tension in these large groupings of elements that the least disturbance sends them flying apart. If mercuric chloride, for example, is mixed with ammonia under certain conditions, a highly explosive compound (Hg_2NOH) is produced by way of the so-called Millon base. Similar compounds result from the chemical reaction of mercury with nitric acid and alcohol. The product is mercuric fulminate ($HgC_2N_2O_2$), an association of dissimilar and conflicting substances which explodes at the slightest tap. This makes it suitable for use in fuse caps for detonating explosives. An interesting point is that nitrogen, though certainly essential to the explosion, lacks the power of combining these heterogeneous elements. For that, mercury's associative capacity is needed.

The god Mercury was the divine messenger whose mediation connected earth with heaven. He was also the god of commerce, thereby bringing men into association with one another. This, too, can be harmful in excess. Nowadays there is a tendency to form large cartels and amalgamations on all sides. Companies, so aptly termed 'sociétés anonymes' by the French, tend to lose any personal character when they expand into huge concerns and giant trusts. Excessive amalgamation dissolves the personal element.

When 'iron will' and the 'mercurial temperament' work harmoniously together, they produce capable and well-rounded personalities with social gifts. But imagine an excessively mercurial temperament, with the capacity to organize and manipulate large combines, and dominated as well by an egotistic will.

What a perfect constellation for producing a Kreuger or a Stavisky!

The god Mercury represented qualities which, in their finest flowering, make for the building of true human community, but they can also lead to an Armageddon if misused. And these divine mercurial qualities in human nature are perfectly mirrored in the physical and chemical properties of quicksilver.

The mercury process is a very important one in the human organism. Any and every 'amalgamation' process that goes on there can be looked upon as mercurial. The term, of course, is used here to designate a force – a force which, carried to the ultimate point in nature, produces the metal quicksilver.

Two important functions of the human organism are easily recognizable as mercury processes. One is digestion. After nutrients have been broken down to a considerable degree by secretions of the mouth, stomach and intestines, the homogenized mass is absorbed through the intestinal wall into the lymph and blood stream. In the course of this, an amalgamation of external nature and the human self takes place. Digestion is thus really a fusion of two spheres of energy which merge in the human body-building process.

The second process takes place in the breathing organs, where human nature and external nature also come together. Through the air we breathe we share in the atmospheric life of the whole earth. All humanity breathes the same air, which enwraps all nature. This fact can be recognized as the true basis of a feeling for an understanding of nature, as well as of the communal sense that breeds truly social conditions.

Even the build of the lungs is mercurial. They are like little trees made of tiny hollow drops. Innumerable spherical spaces thus form a common breathing surface.

We see the imprint of the mercury process on the organism wherever glands built of cells function together as described. Separation into droplets, out of which functional wholes are built is the signature of the mercury forces everywhere. On the human physiological level they control the life of cells. Cellula liquefaction and proliferation are both due to disturbances of the mercury process, which normally maintains a state of balance between the total human organism and its single cells.

Some of the therapeutic uses that can be made of high dilutions of quicksilver will perhaps be obvious in the light of the above account.

CHAPTER THIRTY-ONE

Silver

In many of its properties, silver is just the opposite of lead. Lead's resonance is as dull as its lustre. It melts easily, can only be cast, and is a poor conductor. Silver gives out the most ringing of metallic tones and gleams most brilliantly. It is very subtly workable, melts only at high temperatures, and has the highest degree of conductivity. The two metals occupy opposite ends of the table of metals, like Saturn and Moon in the planetary order. The moon's speedy orbit and ever-changing path give silver an inner mobility. But silver manifests this trait quite differently from mercury, as will be seen in the course of this chapter.

The chemistry of silver paints the first strokes of its portrait in the striking phenomenon of its mirroring capacity. Everyone who has stored silver nitrate has certainly noticed a very fine film of silver appearing on the glass walls of the container. The coating gradually grows thicker, until finally it forms a real mirror. This precipitation process can be hastened by various reducing chemicals. The tendency to form mirroring surfaces is one of silver's chief characteristics.

The chemical reactions of silver show the same tendency. It reacts, for example, with chlorine to form white silver chloride. Under the influence of light it throws off the chlorine with the same energy with which it first attracted it, returning to its former pure metallic condition. Finely distributed, it now appears black. This process is the basis of photography. Photographic plates are coated with an emulsion of white silver chloride, which is sensitive to light. When light streams through the camera lens on to the coated plate, the illuminated parts react, while the rest stays unchanged. The process is completed by developing, which gives us the negative with its black (silver) outlines. The parts that were illuminated are now black, while those untouched by light are still covered with silver chloride in its unchanged white. The fixing bath which the negative is now given removes the silver chloride, so that the dark objects on the plate seem transparent. (A solution of sodium thiosulphate, which dissolves silver chloride, is used for fixing.)

To get a positive print from the negative, the process is run through again. Printing paper is now exposed to light, acting through the negative, and given the same further treatment as the first plate.

This, then, is how photographs are made: they could be called mirrors of the past. A mirror process is certainly also involved in their manufacture.

The Liesegang ring phenomenon helps to round out our picture of silver and of the inner mobility that accounts for its reproductive power. When a drop of silver nitrate falls on a glass plate coated with chrome gelatine that has not quite hardened, the silver reacts with the chrome. A round reddish-brown spot of silver chromate appears. It spreads in all directions, not in the even way an inkspot does, but in wave after wave, each one of which makes a concentric red-brown ring around the original spot. What is characteristic here is the rhythmic repetition that forms concentric spheres, where one might have expected to see just a single sphere as the spot spreads out. There is an outflowing motion with a rhythmical wave impulse, like the spread of a musically vibrating sound. This is another example of the kinship between chemical forces and music; the chemistry of a substance is like an inner music that organizes matter into ordered patterns.

The Liesegang rings recall the concentric ripple patterns that spread out in rhythmically expanding waves from the place where a stone is thrown into still water. We might call both reproductive.

When an object is reflected in a mirror, we speak of a pictorial reproduction. We also speak of photographic reproduction. When we stand between mirrors we see countless reflections of ourselves, very like the concentric silver chromate rings in the chrome gelatine. Ceaseless repetition and wave-like reproduction of some motion or condition of matter are characteristic of the silver force.

Reproduction means, in a narrower sense, the renewal of species in the world of nature. Just as the silver reaction described above spreads and spreads in concentric circles, nature brings forth cycle after cycle of budding, germinating life, as species reproduce their kind. And even in the single organism the same living rhythm of growth goes on. The annual rings seen on cross-sections of tree trunks are an expression of the same force that makes Liesegang rings. Grains of starch seen under a microscope or the cross-section of an egg reveal the same outflowing rhythm.

The silver process is the force responsible for all these life-rhythms. We are referring here, of course, to the action of a universal force that finds material expression here on earth in the substance silver.

In this connection it is natural to find silver tending more than any other metal to the colloidal state. A silver salt need only be treated with a protein solution to produce pure colloidal silver. We know that a colloidal substance can be described as being neither completely fluid nor yet solid. It has a potentiality for either state. This is an essential characteristic of the living. Our blood, plant sap, and all other fluid carriers of vegetative functions, are colloidal in character.

These silver forces are active in all growth and body-building processes in the human organism, and most strongly, of course, where physical life is reproduced: in the sexual organs. The silver process works on a higher plane in the brain, the organ that enables us to reproduce thoughts and to mirror the world in our conceptual life.

The properties of silver as an earthly condensation of moon forces indicate its therapeutic uses. It proves valuable as a medicament in the treatment of regenerative and reproductive disturbances, providing valuable support for recreative functions and permeating the fluid organisms with its vital rhythms. Its many-sidedness leads to its use in a great variety of other special therapeutic applications.

Just as lead brings the forces of Saturn to manifestation, so the silver process reflects the action of the moon, whose immediate influence on all the rhythms in man and nature is everywhere evident.

The rhythm of the tides is currently regarded as a product mainly of lunar gravitation. This could be correct, although the assumption leaves some difficult problems unresolved. In any case, there can be no doubt that the primary cause of tidal ebb and flow is the moon's law of rhythm, noted in all manifestations of the silver process studied in these pages. We know that sea water contains silver in the ratio of 10 mg. to a cubic metre. It may be that silver acts as a focus for lunar influences, and as a medium for transmitting moon forces and their rhythms to the tides. And just as sea water rises and falls in accordance with the laws of lunar rhythm, so is there a tide of sap in plants, both during germination and in later growth.

Plant growth is accompanied by a rhythmic increase and decrease of substances, a tidal emergence and disappearance of matter subject to a monthly rhythm. Certain disturbed states of consciousness, such as epilepsy and somnambulism, worsen or improve with changes of the moon.

In some regions there are still farmers who cling to old traditions and regulate the times of planting, harvesting and other such farming activities by the moon phase. This could well be called superstition, if the method were not proved right by the author's own observations and experiments. Modern man should not depend on old traditions, but should rather investigate the laws of life anew and then act rationally in accordance with them.

Now we might call silver a dense form of moonlight. It is a substance very like the moon in brilliance and reflecting power. The moon reflects the light of the whole universe. Starlight as well as sunlight comes to us in moon reflection. We need hardly be surprised that this dark satellite reflects the sun most obviously. Like a true mirror, it always turns the same side toward us. Its surface is very like that of a frozen flow of silver which has 'spluttered' while getting rid of the quantities of air it absorbed while in a white-hot state, and comes finally to resemble a moon landscape sown with craters.

In ancient times the moon forces manifested in these phenomena were felt to be divine. Greek mythology attributes them to Diana, or Artemis. One of her chief shrines was at Ephesus. The statue of the goddess there shows every attribute of fertility.

The pupils of this shrine knew that the goddess ruled the cycle of the moon and the various vegetative and reproductive processes in man and nature. Celebrating their rites by the light of the full moon, they felt forces of emergent life streaming down to earth. The forces of the new moon, on the other hand, were felt as a stimulus to creative powers of soul and spirit. In glorious visions they beheld the evolution of the aeons out of the creative spirit of the universe.

Somnambulism is a strange, morbid modern echo of these old experiences. Sleepwalkers go on highly dangerous expeditions at full moon, as though their bodies were not subject to gravity. They even seem to be drawn up out of the darkness of the earth exactly as the plant world is, during the full moon phase.

At new moon, moonstruck individuals stay in bed, but often have astonishing cosmic visions in their sleep, albeit mostly in distorted and fantastic forms. We see in these abnormal cases the tendency which moon forces have to alternate between physical and psychic influences.

In Greek mythology this latter aspect is represented in the virgin Diana or Artemis, the twin sister of Apollo, who – in opposition to the Diana of Ephesus – is not portrayed as being physically fertile. Hers is rather psychic and spiritual creativeness, which is intensified when the moon is new. Sometimes she is shown with a new moon crescent on her youthful brow. Unlike Athene, who sprang in full armour from the head of Zeus, she does not picture the austere, illumined life of thought and a striving for wisdom, but is rather related as goddess of the moon to night, with its greater depth of feeling, its creative fantasy.

In the chemistry of olden times, Diana was the name for silver. And just as this goddess has two aspects, silver is found in nature in two different forms: in knobs reminiscent of grape clusters, and in fine hair-like threads, sometimes known as 'silver curls'.

CHAPTER THIRTY-TWO

Gold

Gold occupies the middle position in the list of metals. Since it is the sun metal, we may expect to find it possessing all the contrasting properties of other metals in a harmonious and balanced form. As the sovereign and highest expression of the world of metals, it also exhibits unique capacities.

Gold is most often found in a pure state, usually in a quartz matrix. It almost invariably occurs as a 'pollutant' – if one may call it that – of pyrite. Pyrite, with its golden-glinting five-sided dodecahedrons, always has some gold in it, and its very form reflects the twelvefold rhythm of the sun's course through the Zodiac. It is an interesting and significant fact that gold always occurs very close to the surface of the ground. Gold miners know that they find less gold, the deeper they tunnel. Atmospheric, hydrospheric and geospheric action sometimes wears away the parent rock, exposing the gold. That is why it occurs in the sand of rivers, seas and deserts. The Egyptians, for example, got their gold from the Sahara and the Nubian Desert.

Gold is recovered by reducing the parent rock to sand and then separating out the grains of gold in the age-old process known as washing. This method, whether applied in its most primitive form by the gold-panner or in great technical perfection, rests on the simple fact that gold is heavier than the ground matrix, so that the latter can easily be carried off by a stream of running water. Chemical recovery by means of chlorination and amalgam processes is a thing of the very recent past.

The amalgam process makes use of the ability quicksilver has to dissolve metals by forming amalgams: gold is simply dissolved out of the rock by quicksilver. In the chlorination process, chlorine performs the extraction. Chlorine is the only chemical agent other than cyanide capable of having an effect on gold.

The power gold has to resist virtually every sort of chemical attack entitles it to be ranked among the precious metals. Only a special mixture of concentrated hydrochloric and nitric acid, such as produces nascent chlorine, has any effect on gold, converting it into water-soluble gold chloride. This hydrochloric and nitric acid mixture has long been known by the very suitable name of 'aqua regia' (royal water).

To call gold the 'king of metals' is to use a figure of speech that means little to the modern chemist. But we shall see that a closer study of gold and its less obvious qualities proves its right to this exalted title.

Gold occupies the central place in the table of metals arranged with reference to such dynamic properties as lustre, resonance, conductivity and malleability. Like the sun, which is the harmonizing centre of the heavens, governing and ordering the curving paths of the planets, gold is an expression of the harmonizing force in man and nature. Gold's aristocratic nature keeps it free of entanglements and enables it to mediate between extremes. If gold were to form chemical bonds with all sorts of other substances, it could not rule over them as king.

How impressively the changes of colour which gold undergoes in its various metamorphoses reveal its all-embracing, universal nature! The metallic gold we are familiar with in ornaments and tableware has a warm lustre, like a sunny late-afternoon in summer-time. When it is hammered into paper-thin gold-leaf, however, it becomes translucent and turns a glorious emerald green against the light.

Goethe's theory of colour can help us understand what this phenomenon signifies. Goethe pointed out that pure yellow and pure blue are the colours most closely approximating to light and darkness, and are thus polar opposites. Red and violet are intensifications of these poles. Green is their harmonizing. The earth's green vegetation which the sun conjures forth is the harmony of light and darkness, sky and earth, in living matter. This could explain why gold occurs right at the surface of the ground, where earth and cosmos are evenly balanced. Summer's glorious display of green foliage seems conjured forth in light's transformation by the sun-gold process. Everyone knows what a soothing effect green has, both out of doors and in green-painted rooms. Green stands calmly poised between red's aggressiveness and the nostalgic yearning, the solemnity, the exaltation of blue-violet.

But there is another colour that creates a harmony between red-yellow and blue-violet: purple. This is not a simple mixture of two polar opposites, but a synthesis on a higher level. Purple may be called a higher metamorphosis of green. Green sustains and carries us. We stand on the earth's green surface and find security and peace of soul in green. It conveys a sense of earthly balance. We can experience the polarity between red-yellow, with its glowing activity, on the one hand, and on the other the quiet remoteness and sublimity of blue-violet.

Over this lively interplay of colour, purple reigns supreme. There is scarcely a human being who does not sense the majesty of this indescribable colour. From earliest times it has been used in the trappings of dignitaries as a mark of distinction and high office. Kings and priests, as leaders through whom the divine will spoke to humanity in times gone by, were robed in purple. Purple seems to open a door through which a higher world can enter human souls. But the purple we see with physical eyes is little more than a dark shadow of the true colour, which Goethe called the heavenly child of the Elohim. He says: 'A great secret begins to reveal itself to us when we rightly conceive the moving apart of blue and yellow, giving special attention and appreciation to the intensification towards red, for opposites here bend back towards each other, uniting in a third hue. The spiritual import of these two separate and opposite beings begins to dawn on us. And as we see them bring forth green below and red above' (in our designation, purple) 'we shall scarcely be able to resist the thought that below we behold an earthly colour; above, the heavenly offspring of the Elohim.'

This pure purple is the colour gold assumes in the colloidal state. The reduction of a watery solution of gold chloride makes it possible to produce every shade of purple. The more it is diluted, the more delicate and glorious the purple grows. A colloidal solution of one part gold in ten million parts of water has a marvellously subtle purple tinge that approaches peach-blossom and is like the indescribable bloom in healthy children's cheeks.

Gold-purple was used in the Middle Ages for staining glass. The glorious purples that shine with all the magic of a long-forgotten art from old church windows, such as those at Chartres, were made of this gold-purple. Present day glass-stains derived from gold are not to be compared with the medieval product.

The world-embracing power at the heart of the universe was experienced by the ancients as the Sun-Spirit. Zarathustra and his followers called this being Ahura Mazdao, the great aura of the sun. Initiates of olden times knew that in the course of evolution the Sun-Spirit would leave the sun and link himself with the destiny of the earth. It is significant that the Egyptians mourned their sun-god Osiris, for they were no longer able to behold him in the sun. This fact was symbolized in the mythological picture of the dismemberment of Osiris by his brother Typhon. The Greeks venerated in Apollo only one aspect of the Sun-god, whose spirit had meanwhile drawn closer to the earth. This explains why Zeus, or Jupiter, the father of Apollo, reigned on Olympus in his stead. And the sun-hero Baldur also passed from the consciousness of the Teutons; the twilight of the gods set in because man could no longer reach the inspiring forces of the universe in the world outside him. The Sun-god incarnated on the earth as Christ. This event was experienced by Celtic and Germanic initiates in their Mysteries, as well as by the few people who witnessed it in Asia Minor. Even the Finnish Kalevala contains a reference to the Son of the Virgin. At this turning point in human history there took place a great transformation, whereby the cosmic forces that had guided man became the innermost powers of his soul – conscience and reason, the basis of capacities to be developed in the future. The Teutons, especially, felt a conviction that the darkness of the twilight of the gods would be followed by a new epoch. Widar stamps out the power of the Fenris-wolf with a shoe made from pieces of leather given him by men who had worked together. Out of work, and particularly work in community, will flow the forces for remaking the earth. Can we conceive a more magnificent vision of man's future capacities?

Such forces spring from the heart, the central organ of the human body. Here is to be found the wellspring of all harmonizing currents, the strength that balances polarities, the fount of healthy social feelings. The heart may be destined to play a more important part in future developments of consciousness than we can imagine today. Wisdom that is golden – in other words, imbued with the forces of the heart – will be needed to bring the affairs of earth into balance and harmony.

As the sun affects the development of physical life, so has gold affected the history of man. In ancient times gold was held sacred, for it was looked upon as belonging to the highest gods. In ancient Egypt, the private ownership of gold was forbidden. Gold was kept in the temples, and priests and kings as earthly representatives of gods carried it during ritual celebrations.

Not long ago a gold helmet that is probably the oldest golden treasure ever found was discovered in Ur of the Chaldees. A vehement controversy arose over the possible uses of this helmet. After what has been said above about the character of gold, it should be clear that the helmet must have been a ritual object, not part of a warrior's armour. It carries a head-encircling sunlike disc, probably meant as an image of the aura which indicates inspiration from the macrocosm.

Later on, when men began to hanker after the ownership of gold, it was gradually degraded to a symbol of personal wealth; the full extent of the curse that accompanied the plunge into egotism fell upon it. Things have now gone so far that all the gold has disappeared into the vaults of national banks, and pieces of paper are used instead of it. Is there not a striking parallel here with the gradual change from divinely guided consciousness to abstract individual thoughts, which are often as far removed from reality as paper is from gold?

Gold will be freed from these fetters of expediency to the degree that mankind works its way out of abstract thought to a new and active consciousness. We saw the first steps being taken in this direction when the gold standard was abandoned and currency was based instead on real values created by human industry and effort.

The properties of gold described above are also apparent in its use as a medicament when the various physiological functions have got out of balance. It is especially valuable for disorders of the heart and circulation.

Gold reminds us of the aluminium process, which it resembles in its capacity to harmonize polarities. The quintessence of forces of harmony exhibited by aluminium in precious stones appears again in the precious metal gold, but in closer relationship to us.

Special reference was made above to one gemstone, tourmaline, which is seen in cross-section to have both green and purple colours. These are the colours that reveal gold's all-embracing character in the translucent green of gold-leaf and in gold-purple. A tourmaline set in gold thus comes to seem especially symbolic of harmony between earth and universe.

CHAPTER THIRTY-THREE
The Brothers of Iron

COBALT, NICKEL, CHROMIUM, MANGANESE,
TUNGSTEN, VANADIUM, PLATINUM

We have described the seven principal metals, but have not yet mentioned a whole family of others which we all use daily. The first thing that strikes us about metals such as cobalt, nickel, chromium, manganese and the like is their relationship to iron. They are all found in association with it. Most iron ores, and so-called fahl-ores in particular, invariably contain one or more of these collateral metals. It is not always easy for chemists to identify them in a compound, so closely related are they chemically.

When we discussed the great difference between metals and non-metals, we spoke of the Periodic System, in which no proper place can be found for metals. The metals of the eighth group, to which quite a few of these collateral metals belong, are particularly isolated; they are set off by themselves, with no relationship to the system as a whole. It is not even decided whether nickel should precede or follow cobalt in the list, as both have the same atomic weight. The Periodic System would undoubtedly gain in clarity and realism if the metallic elements were eliminated, as in the chart below. This would leave seven, instead of eleven, series in the system, and reduce its eight groups to seven.

The only metals which this arrangement would leave in group five are arsenic, antimony and bismuth. These are neither true metals nor true earths, but their properties belong under the heading of Group Five. We shall have more to say about them in Chapter Thirty-four.

The connection between metals of the iron group is further demonstrated by the relationship some of them have to magnetism. Cobalt and nickel both behave like iron in a magnetic field.

The relation of all these metals of the iron group to carbon is also like iron's in forming hard, brittle carbides that dissolve in the molten metal. In the manufacture of steel, quite small additions of these collateral metals are enough to produce varieties of

ROW	GROUP I	II	III	IV	V	VI	VII	VIII
1	H I							
2	Li 7	Be 9	B II	C 12	N 14	O 16	F 19	
3	Na 23	Mg 24	Al 27	Si 28	P 31	S 32	Cl 35,5	
4	K 39	Ca 40	Sc 45	Ti 48	51 V	52 Cr	55 Mn	56 Fe 59 Ni 59 Co
5	64 Cu	65 Zn	70 Ga	73 Ge	As 75	Se 79	Br 80	
6	Rb 85	Sr 88	Y 89	Zr 91	94 Nb	96 Mo	99 Tc	101 Ru 103 Rh 106 Pd
7	108 Ag	112 Cd	1151 n	119 Sn	Sb 122	Te 128	I 127	
8	Cs 133	Ba 137	La 139 / Lu 175	Ce 179	181 Ta	184 W	186 Re	192 Ir 195 Pt
9	197 Au	201 Hg	204 Tl	207 Pb	Bi 209	Po 210	At 211	
10	Fr 223	Ra 226	Ac 227 / Lw 257					
11								

Fig. 31

Periodic table of the elements (omitting the noble gases, Group O).

steel with special qualities of hardness, toughness and elasticity. Chrome, nickel and tungsten steels and the like are known even to the layman as materials possessing steel's characteristic properties in the highest degree.

Metals of the iron group can be worked like iron; they can be either wrought or cast. They have almost identical qualities of resonance and lustre. They conduct heat and electricity to the same degree. The following table, in which the conductivity figures of this group can be compared with those of the seven main metals, illustrates this plainly:

	Conductivity of			
	Warmth	Electricity		Speed
Silver	100	100	Moon	392
Copper	74	77	Venus	32
Mercury	(68)	(76)	Mercury	36
Gold	53	73	Sun	30
Iron	17	17		
Cobalt	17	17		
Nickel	17	17	Mars	18
Chrome	18	19		
Manganese	17	20		
Platinum	17	16		
Tin	14	13	Jupiter	4
Lead	8	10	Saturn	2

These conductivity figures make it plain that we have to distinguish several secondary metals from the seven principal ones, and that the former are connected with iron as though by family ties.

Now why does iron have so many brothers?

It has been shown that the iron forces work inward from the periphery of the cosmos towards the centre and are agents of the densification that prevails on earth. We came to know the Mars process as the force that makes incarnating possible. Thus there must always have been a stronger than usual bond between the earth and Mars, even before the earth became a solid body. If this is true, we might suppose that primal Mars substance was deposited on the earth when the paths of the two planets crossed.

This Mars substance is not to be imagined as being in the present-day material state, but in the much finer one suggested by our term 'iron process'. This substance must have been thoroughly exposed to the various forces of the earth, which would then have brought about a variety of changes in it. We will try to read the nature of these changes from a study of their products, the metals we have called 'brothers of iron'.

COBALT

A study of cobalt suggests that it is an even earthier form of iron than iron itself. Like iron, it reacts to magnetism, forms carbides with carbon, and has the same chemical reactions. But it has other aspects which indicate that its bonds with the earth are stronger.

Cobalt's ores and salts are both tinged with a dark and melancholy shade of blue-violet. It has less affinity to water than iron has. Red cobalt salts give off their water of crystallization on contact with dry air and turn dark blue. This makes it possible to use a solution of cobalt salts for invisible ink. When the paper is warmed, the writing turns a visible blue. Cobalt salts are also used to make weather-indicator mannikins. Their treated clothes turn a darker blue, the drier the air becomes.

While cobalt salts are chemically more mobile than the salts of iron, cobalt is more resistant as a metal. Iron objects such as the bonnets of motorcars can therefore be protected with a thin cobalt plating.

Now what was the origin of cobalt?

In the beginning, before the earth became a dense material planet, it teemed with flowing currents of awakening life. Men of olden times experienced this non-embodied life as a vivid world of many sorts of nature-spirits. The mythology of the Greeks and more especially of the Celts and Teutons is full of accounts of beings that rule over nature from behind the scenes. German fairy tales are another example. They are not to be lightly dismissed as primitive entertainment for the very young and undeveloped, but are rather picturizations of real facts. They take us into the realm of gnomes or kobolds, water-sprites or nixies, elves or air-sprites, and fire-beings, also known as salamanders. These are the beings whom men saw in pictorial vision weaving the tapestry of nature.

We moderns are inclined to relegate these fairy-tale creatures to the realm of fantasy or to call them personifications of forces at work in the natural world. Is it not at least quite as logical to speak of forces as expressions of the activity of nature-spirits? Shepherds, peasants and woodmen who live in closest touch with nature (especially in far northern lands, where human beings alone with nature and undisturbed by such products of civilization as the radio and cinema, preserve deep-going bonds with nature and real simplicity) still testify in many ways to the reality of these legends. Miners and foundrymen of the Middle Ages were very familiar with the problems caused by secondary metals in the forging of iron, and in their simplicity (or wisdom) attributed these to the interference or activity of kobolds and nixies. Cobalt and nickel are two of the metals named after this association.

Cobalt, then, really means kobold. The kobolds that gave the metal its name are also known as gnomes, dwarfs and earth-spirits, and are always pictured as having to do with the earthy element and living deep inside the earth. They are said to work veins of ore, collect precious stones and metals, chip away at rocks, prepare the soil for plants and work around their roots. They are pictured as very clever fellows, especially at calculating and reckoning. Sometimes they tease and torment human beings, but often they help them. Of melancholy temperament, they yearn most movingly for daylight. Little red jackets are their special joy.

Cobalt may thus be termed an iron metal with a gloomy, melancholic tendency, even more deeply related to the earth than iron is – in short, iron that has gone beyond itself.

NICKEL

Nickel is like iron that has gone further in a nixie-direction; in other words, has come under the influence of water-sprites. Even the quality of its lustre recalls the reflection from a watery surface.

All the nickel salts are a glorious deep-sea green. Like copper salts, they contain a large amount of water of crystallization, and can easily be dissolved into beautiful green fluids that make one think of deep Alpine lakes.

The similarity to copper is found also in nickel's tendency to form complex salts. The addition of ammonia to nickel-salt solutions produces a green nickel hydroxide precipitate. A further addition of ammonia dissolves this again into a glorious sapphire blue liquid. There are other brothers of iron capable of forming similar complex salts, but with nickel the colour aspect is so outstanding that one is reminded of the Venus nature. Again, both nickel and copper are ductile.

Nickel takes its name from the water-sprites called nixies, or sometimes undines. These beings are described as the guardian spirits who watch over rivers, lakes, ponds and springs. They are active also in the green sap of plants where they stimulate nature's growth processes.

Nevertheless, nickel is a true member of the iron family. Like iron, it is magnetic. It dissolves carbons, forming carbides. These, alloyed with iron, make nickel-steel, a particularly tough metal used in the manufacture of gun barrels and other articles where toughness and resilience are prime requisites. Nickel-steel is resilient rather than hard; it can be stretched to twice its normal length without breaking apart.

Nickel shares with cobalt a greater degree of resistance to chemical corrosion than iron has. A coat of nickel-plating makes the surfaces of metal objects relatively invulnerable. We shall find this property increasingly enhanced as we go through the list of iron-metals.

The process wherein nickel is used for hardening oils rests on a similar property. Fluid oils are changed into solid fats by using very finely ground nickel. The nickel itself is not affected by the reaction; it merely supplies the necessary chemical energy. Processes of this kind are called catalytic and the agent is called a catalyst. This property, too, becomes increasingly marked as we go through the list of iron-metals from nickel to chrome, manganese and platinum.

The general impression given by nickel is one of a Mars substance that has taken on something of the copper nature through its relation to water-spirits, always servants of Venus. Its catalytic power shows that it has considerable free chemical activity; but its other qualities, as a member of the iron group, it has taken a further step down towards hardening.

CHROMIUM

Chromium, too, reveals its kinship to the iron family. It is always found in conjunction with iron. Chromite (chromic iron ore) is the most important source of chromium.

Like iron, chromium dissolves carbon and combines with it to form carbides. Alloyed with iron it produces the chrome-steel famous for its hardness. Thus it represents another downward step towards sclerotization.

The Jupiter influence is more marked in chromium than it is in iron, in the sense that certain tin qualities are more in evidence. We have described how tin organizes light and air, conjuring forth latent colour as it does so. Tin is thus used as a mordant in dyeing wool and silk. Chromium has the same capacity. Where iron makes only a modest showing as a mordant, chromium has everything required. It plays an important part in madder or alizarin dyes. The colours it produces have a somewhat darker tone than those produced with tin mordants. This is only to be expected, since chromium is a relative of iron and hence tends to iron's earthy qualities. This may explain why we find in chromium the fixed colours of its salts – chrome yellow, chrome red, and so on.

Cobalt was pictured as the result of the working of gnomes on iron, nickel the result of a similar activity by water-beings. Chromium can be recognized as iron that has undergone a transformation at the hands of sprites who are at home in the airy element. They are called sylphs or elves in fairy-stories. There we find descriptions of how these beings live in wind and weather, in the streaming currents of the atmosphere while accompanying birds in flight. The bird of Jupiter, the kingly eagle, was said in olden times to be seeking out its master's throne in its flights through the sunlit realm of clouds. In the sense that water-sprites work hand in hand with the Venus forces active in the watery element, air sprites or elves may be called attendants of Jupiter.

The properties of chromium make it suited for use in tanning, which we have called a mummification process. The rapid tanning that can be carried out with chromium shows how particularly prone it is to atrophic states.

At first glance, this aspect of chromium seems quite out of keeping with the airy lightness of bird flight. But the same opposite aspects are to be found in birds themselves, where the sclerotic tendency reaches a high point, while at the same time they are master of the air's free realm. A bird's foot is scarcely more than skin and bone, and its whole organism is atrophic. The process that in man effects the head only is exemplified in the whole bird: it is essentially a head-creature, with its digestive system as a kind of appendage. The human head develops consciousness at the expense of vegetative processes; in the same way the bird develops flight and song.

Chemically too, chromium shows more leaning to atrophy than either nickel or cobalt does. This means that it is more resistant to invasion by chemical agents and the atmosphere. Metal objects can be protected against both by a chromium plating.

MANGANESE

Manganese is the 'magnesium nigrum' spoken of by Pliny. Probably he used this term to call attention to its fiery nature. Its ore is invariably found in company with iron; it has iron's relationship to carbon and makes a very hard manganese steel. Apart from these iron qualities, it has a whole series of other noteworthy characteristics which relate it strongly to the fiery element. Its salts range from rose-red to a flaming violet. Finely ground braunite, a manganese dioxide occurring in nature, causes alcohol and ether vapours to ignite. Manganese dioxide, which in the form of pyrolusite is the commonest and most important of manganese ores, is an essential of glass manufacture. Stained and impure glass becomes clear and bright when smelted with finely powdered pyrolusite. It is as though fire's cleansing power were active in it. Indeed, pyrolusite means 'fire-washer'. This substance, which certainly looks externally like a victim of sclerosis in its resemblance to pieces of burned-out slag, and yet on the other hand has so strong a fire-force inherent in it, reveals a decidedly Saturnian nature.

The sclerotic tendency of manganese comes out in still another interesting phenomenon. Linseed oil is used in the manufacture of oil-based paints because it resinifies in time and forms a hard surface. It is therefore known as drying oil. Poppy-seed oil, used in making the more delicate artists' paints, is in the same category. A surface painted with linseed oil alone takes weeks or even months to dry. The addition of manganese in the form of fatty acid salts reduces the drying time to a matter of days or hours. Manganese compounds are therefore called siccatives, or quick-driers, and it is interesting that when added to colours they give them an especially warm lustre.

In the animal kingdom, insects show the same proneness to desiccation, together with the same relationship to warmth. Sclerotization can go no farther than it does in insects. Bees, wasps, hornets, and the various beetle species are completely dried-out forms; yet they have a special relationship to the warmth element as it lives in air, in nature and in plants. Bees are related to the blossom, wasps and hornets more to the fruit element. Blossom and fruit are the parts of the plant in which cosmic warmth processes reach culmination. And there is a latent but fiercer fire in the poisonous stings with which these insects are endowed.

The ancients saw the activity of fire-spirits as the underlying reality of these fire-processes. In fairy-tales and legends these spirits are linked with blossoming, ripening and fruiting, as well as with hospitable hearth-fires. One feels they might have touched even as hardened a substance as manganese with their fire-fingers.

TUNGSTEN

With each further step in our study of the iron family we come upon metals with increasingly marked hardening properties. Tungsten, for example, began quite recently to play an important role in the iron and steel industry. Alloyed with iron, it makes a steel almost as hard as diamond. Unlike ordinary steel, it stays hard even when heated to the red-hot point. This has made it indispensable in steel-working tools. These qualities have earned tungsten steel the name of 'noble steel' (*Edelstahl*).

A related characteristic of tungsten is its invulnerability to corrosion by acids, even aqua regia – as though it were a precious metal.

Tungsten glows, on being heated, with a beautiful white light. Because of its resistance to chemical change it has long been used in the manufacture of electric light filaments.

VANADIUM

Vanadium's relatedness to iron appears in many of its properties. It always occurs in company with iron ores; small quantities of it are found especially in the so-called 'pea ores' (oölithic limonite). Strange to say, traces of it are often found in cultivated soils and in the ash of grapes and oak-trees.

Like iron, again, vanadium has an affinity to carbon and forms silvery white carbides. It is like tungsten in producing a steel almost as hard as diamonds when alloyed with iron, and in being used to make machine tools. Vanadium steel also ranks among the 'noble steels'.

PLATINUM

Platinum gets its name from the Spanish 'plata', meaning silver, and one can indeed regard it as a dead form of iron imbued with silver properties. Molten platinum spatters on cooling just as silver does. When in a molten state it absorbs air, which it expels again on solidifying. But its most outstanding characteristics are its non-corrosiveness and its catalysing action.

Anyone who has had to work with platinum containers – when heating substances in platinum crucibles for instance – will remember how carefully he had to protect the crucible from the bright (carbon-containing) part of the gas flame, for it becomes brittle and fragile on exposure to it. This is due to the heated metal's readiness to combine with carbon. Platinum is capable of only one other chemical change: it corrodes when brought into contact with aqua regia. The carbide-forming tendency of platinum is slightly reminiscent of its relationship to iron.

Platinum is not only a first-rate catalyst – it is *the* catalyst par excellence. Its most familiar role in industry is in the contact process of manufacturing sulphuric acid. Sulphur becomes sulphur dioxide on burning. To oxidize it further and make sulphur trioxide used to mean putting it through complex and elaborate processes such as the lead chamber process. Ever since the catalytic effect of platinum became known, platinum has been used, finely ground, as a contact material. Now oxidation of sulphur in the manufacture of sulphur trioxide proceeds simply and smoothly, leaving platinum itself quite unaffected by the reaction.

A further reaction which platinum's catalysing action speeds up considerably is that of hydrogen and oxygen. So stimulating is this action that spontaneous combustion follows. Those who go back to the gas mantle era will remember the lighters that used to be laid on the glass chimneys. A minute quantity of finely ground platinum was sprinkled over the lighters, and the gas ignited by itself on contact.

Many other reactions are set off or speeded up with platinum as the catalyst. This is especially so with organic compounds.

While on the subject of platinum as the outstanding catalytic agent, it may be well to pause for a closer look at what catalytic action really is. It has been mentioned that nickel, used as a catalyst to harden oils, holds back from active material involvement in the chemical process taking place, and simply supplies the chemical energy that nickel radiates, so to speak. This radiation is a feature of what we might call the onset of sclerosis in the substance itself.

We can get a better understanding of this process by studying comparable situations found in nature's higher kingdoms. The same polar development takes place, for example, in the nerves. Nerves have a constant tendency to become sclerotic. They are always in a dying state, thus freeing the life that is very loosely attached to them. The life of thought and sense-perception is built on this radiation of forces freed from the body. Nerves thus present a polar contrast to blood, which is life-saturated and wholly involved in the physiological processes and organic functions of our material bodies. We also find a reflection of this dying process in the nerves in the animal kingdom; it was discussed in relation to birds and insects. The more atrophied and sclerotic the body of an animal, the more freely the species-intelligence functions. The animal intelligence active here is of the species as a whole, not of the single creature. Bird migrations and the organization of beehives and ant heaps are astonishing examples. Nobody can really think that the individual swallow, bee or ant possesses such wisdom. These animals are clearly units directed by a superior intelligence operating not from within each animal body but from outside it – an intelligence which supports, directs and organizes the whole community of a flock of migrating birds, an ant heap or a beehive.

This same phenomenon can be seen in the mineral kingdom in the metals of the iron family, where the chemical function withdraws from the metal itself to become active in its immediate environment. We see this happening when nickel hardens fats, when manganese is used as a drying agent to speed up the resinification of oil paints, when it carries the oxidation of alcohol and ether to the point of spontaneous combustion. In tungsten and vanadium the tendency to atrophy becomes increasingly apparent, so that these two make effective catalysts. This development reaches greatest intensification in platinum and the platinum-related metals such as osmium, iridium, palladium, and the like. As a chemical substance, platinum represents a stage of almost total atrophy and sclerosis. It is dead to the point of being unable to participate in any chemical reaction whatsoever. It makes up for this by being enveloped in what we might call an aura of chemical energy that sets off and speeds up many chemical reactions that would otherwise not take place. All the while, the platinum itself remains materially unchanged. We might put it thus: it pays with death for its noble capacity to bestow energy on other substances.

We now understand why platinum is immune to chemical attack and is considered a precious metal. But its aristocracy is of an entirely different order from that of gold, which consists in keeping other substances at a distance. Platinum cannot react with other kinds of matter because it is dead. It belongs in the ranks of precious or aristocratic metals because it has sacrificed its chemical life.

So we see that the iron family is a group of metals that follow the main trend of the Mars metal further into the hardening forces of the earth. As they do so, each brings to the fore the properties of some other planetary sphere: Venus's relationship to water, Jupiter's to air, and Saturn's to fire.

One can, moreover, discern in the brothers of iron the working of a great law, which may be stated thus: Material hardening frees the forces once spent to maintain life in all its suppleness, and they reappear in a changed form as new qualities.

CHAPTER THIRTY-FOUR

The Childhood of Metals

ANTIMONY, ARSENIC AND BISMUTH

Our study of the iron family has enabled us to see in a new light the nature of those metals with which modern times bring us into daily contact. But chemists would say that the entire list of metals has by no means been covered. Antimony, arsenic and bismuth are still absent. They have an unusual and basic connection with the non-metallic minerals, and, as is clear from a look at the Periodic Table, it is hard to know whether they should be described as no longer metals, or not metals yet. The following account will clarify what this means in the various individual cases.

ANTIMONY

Antimony has properties that are in certain respects the exact opposites of those of the iron family. This comes out in comparing iron ores – marcasite and siderite, for example – with antimonite. Antimonite, in the form of antimony pyrites, is the most important antimony ore.

Antimony is diamagnetic. This means that instead of aligning itself, as iron does, with the lines of the earth's magnetic current, an antimony needle points at right angles to it. It is as though antimony were trying to hold magnetism at bay, to get away from it. This is borne out by the following phenomenon. When antimony in solution is electrolytically precipitated, it appears on the electrode in a modified metallic form as so-called explosive antimony, a substance that explodes on exposure to heat or friction.

Both phenomena suggest that antimony does not like being on the earth. As antimonite, it assumes the typically lively radiating form of cosmic forces (cf. fig. 29), and it rejects the dark forces of terrestrial magnetism. When an electric current forces it out of solution and into the form of a metallic precipitate, it reacts with extraordinary violence and wrests itself free of the fetters of this earthly form by exploding.

Hoarfrost, with its myriad tiny points and needles, shows the activity of the antimony-process in the atmosphere. But how does the hoarfrost take shape?

We know that when the skies are clear the air is permeated with transparent water vapour in the finest state it can assume, the gaseous. When sudden cooling of the upper layer causes this fine vapour to condense, it skips the fluid state and assumes the form of countless tiny ice-needles. This is the origin of the feathery cirrus clouds. When this process takes place close to earth, ice-needles form on every convenient point, ridge and slope, precipitating what we call hoarfrost. What is involved here is a sudden rather than a gradual condensation of water vapours – so sudden as to bring about immediate solidifying. This is why hoarfrost is deposited in the ray formation we are familiar with in cyrrhus clouds.

Cumulus clouds form in the middle atmospheric belt. Here, cosmic radiation is balanced by the centripetal formative forces of the earth. This explains why sphere-shaped drops of water are to be found here, rather than clusters of rays. A tendency to spherical forms is the dominant characteristics of the cumulus cloud. When formative forces balance forces of buoyancy so perfectly that neither air nor water gains the upper hand, their interaction produces colloids – not in this case a mixture of the solid and the fluid, but of the fluid and the aeroform. Mixtures of the latter type are called aerosols, the former hydrosols.

If the balance is upset and centripetal formative forces gain the upper hand, the drops of water start to swell and rain ensues; cumulus clouds turn into rain or nimbus clouds.

The last stage of this earthward journeying is reached when rain water comes to rest on the surface of lakes and oceans as a liquid mirror that under suitable conditions can freeze into the rigidity of ice. Ice, in sharpest contrast to hoarfrost, is a solid, homogeneous mass.

In antimony, the same process takes place in the sphere of metals. Antimony seems to be a solidified remnant of the metallity of an earlier period of evolution, when metals were as yet undifferentiated. We might say that antimony was precipitated directly out of an embryonic phase of metallity into earthly form without passing through transitional stages. It is a cosmic infant suddenly transported into earth conditions to which it has not yet had time to adapt itself. But we can gradually adapt it by passing it through chemical and physical processes. We can then discover traces of latent quicksilver, copper and silver characteristics in this young metal, all of which reflect phases of a stepwise descent that antimony by-passed through its sudden precipitation.

Antimony can amalgamate with almost every other metal. In this respect it reminds us of quicksilver. It melts easily and tends to form little drops or globules when poured on to a flat surface, moving in quick parabolic curves with a liveliness again reminiscent of quicksilver. Is there not a parallel here to the way vapour changes into drop-form and builds up cumulus cloud formations? But antimony seems to be trying to get back to the cirrus cloud form, for when its droplets come to rest on the dish they are covered with a sparkling 'hoarfrost' of antimony oxide.

Antimony shows also a decided inclination to enter into very complex chemical bonds. Like copper, it forms all sorts of unusual relationships, especially with tartaric acid and tartaric-acid salts. This comes from the vitality that makes both copper and antimony capable of effecting so many transformations. Antimony's vitalizing forces are well known to medicine. There is a parallel here also to that further phase of the water-cycle where moisture descends from the heights to earth and fructifying rain permeates the ground, collects in lakes and streams, and makes the earth fertile.

The final stage, when water comes to rest and solidifies into a mirror of ice, finds its parallel in the forming of the antimony mirror. Almost all solutions of antimony tend to form mirrors. Colloidal solutions of antimony and of its compounds produce the most wonderful mirrors. The silver process enters in here. But antimony makes it possible also for the radiance of the cosmos to appear in all the colours of the rainbow. Colloidal solutions of antimony compounds, particularly those containing sulphur, coagulate into fiery red, orange and yellow mirrors, from which we get such names as antimony ruby, antimony cinnabar, gold-sulphur, and the like. And we can recognize antimony's tendency to create hoarfrost forms even in these completely solid mirrors, for feathery patterns are often to be seen in them.

This infant, antimony, is always longing for its cosmic home, for a return to the unborn condition. This can be gathered from a further series of phenomena. If, for example, the gas antimony hydride is quickly cooled to, say, 100°C. below freezing point, leaving the hydrogen to oxidize and thus gradually disappear, we get a pure but completely unmetal-like form of antimony. It looks just like sulphur or phosphorus, and is both explosive and capable of spontaneous combustion. The very low temperature of −100°C. has forced it to condense in a region that lies – to keep to our metaphor – even above the cirrus cloud belt.

It is quite in character for antimony to be unable to accommodate itself to the earthly salt-condition. Chloride of antimony is thus a fatty substance, with the fitting name of 'antimony butter'. It even behaves like butter in being soluble in ether and other organic solvents.

We see, in fact, that antimony is only on the threshold of material manifestation as a metal.

These properties of antimony indicate its therapeutic value. In high dilutions it brings the organism youthful vital forces, not indiscriminately proliferating, but penetrated with formative radiations. The fact that it has solidified into this radiating pattern out of an initial undifferentiated stage makes its effect upon the blood understandable. For blood also represents a state of equilibrium similar to that found in the cumulus cloud formation, where solidifying and etherealizing processes are in balance. Antimony regulates this balance, thus becoming a valuable means of treating haemophilia and other such blood diseases where there is a lack of formative, coagulative forces in this liveliest of organs, blood.

ARSENIC

Imagine the youthful qualities of antimony raised to a still higher level, and we have arsenic. Arsenic has no form at all; it exists only as a kind of dust. This is a way of saying that it is extremely dry. It remains a loose powder, and even in compounds it almost never becomes a crystal or a mineral structure with any shape to it.

Arsenic and its compounds normally skip the fluid state and go directly over from the solid to a vaporous or smoke-like one. It is thus always caught up in disintegration, decay, and dusty dissolution. Hence it is almost impossible to find arsenic anywhere in nature as a compact mass; it is usually present only as a trace in other ores, as though particles of cosmic smoke had drifted into them.

Keeping to our picture, we may say that this smoke-aspect of arsenic is like a reflection of the cosmic conditions that obtain in the stratosphere. When the blue sky is veiled by a scarcely noticeable whiteish haze, as happens particularly often in the spring, the haze comes from stratus clouds that consist, not of a compact mass, but of a transparent high-altitude 'smoke'.

It is often a question whether arsenic really is a metal or belongs to some other realm beyond the metallic. Arsenic vapour condenses with quick cooling into a phosphorus-like substance wholly unlike a metal. It is called yellow arsenic, and actually does behave much as phosphorus does. It is soluble in carbon disulphide and other organic solvents, and is so volatile that its penetrating garlic odour carries a long way.

We can perhaps say that arsenic is a substance that marks the evolutionary moment when metals were helped by phosphorus to assume material form. The phosphorus process is, as we have seen, a midwife for many such births, since its activity embraces both the loftiest aspects of the spiritual and the depths of materiality.

Antimony was called a child among metals; in the same sense we must look on arsenic as being at the embryonic stage of metallic development.

Its therapeutic use is in keeping with the characteristics here described. It has a drying effect, removing excess fluids from the organism. Its tonic effect is also understandable; it stimulates the vegetative processes to retain the fluids on which their functioning is based and which arsenic threatens to withdraw. This reaction gives an impetus to the vital forces, but of course only in the proper minimal doses.

BISMUTH

Bismuth is a heavy metal. Unlike arsenic, it is found in its native state in rhombic crystal cubes, with feathery cleavage patterns on their surfaces. Bismuth is extremely brittle and can easily be reduced to a fine powder. Everything about it gives the impression of great age and sclerotic forms. Consonant with this, it is typically found in many cobalt and nickel ores – linked, that is, with members of the iron family.

On the other hand, bismuth also has childlike qualities such as those attributed to antimony. Like antimony, it is diamagnetic, resisting the magnetic field; and in the neighbourhood of such a field it even repels electricity.

Despite its earthly heaviness and its old-man's wrinkles, bismuth too has not yet adapted itself to earth conditions with regard to salt-formation; it is incapable of forming regular salts. When, with the use of a great deal of acid, one finally succeeds in dissolving this metal, the addition of a tiny quantity of water is enough to cause it to precipitate out again in the form of a hydroxide or a basic salt. All solutions of bismuth salts are hydrolysed on coming into contact with water, and a white precipitate results.

One of bismuth's most interesting properties is that it can be alloyed with tin and lead. This alloy, known as Wood's metal, has such a low melting point that warm water melts it.

All this suggests that bismuth belongs to realms of undifferentiated childlikeness, but has been overwhelmed by earth forces which have impressed their sclerotic, old-man qualities upon it. Bismuth is like a child grown old before its time.

Antimony and bismuth are at opposite ends of the thermo-electric series. This means that when the two metals are brought into contact with each other and heated, an electric current flows from the antimony to the bismuth. Atomic physicists say that antimony has more free electrons than any other metal. These flow to any other metal capable of absorbing them when the latter is connected with antimony to form a thermo-electric element. Bismuth is the metal most richly endowed with this capacity. This strange phenomenon indicates how like and yet how opposite bismuth and antimony are.

Bismuth has traces in it of all the metal processes, from Saturn (lead) to the moon (silver), but they are so blurred and criss-crossed that it would be hard to describe them adequately in the space at our disposal.

Certainly it is fair to say that antimony, arsenic and bismuth were once even less solid and less differentiated substances, like quicksilver, before they were caught by the wave of densification responsible for the hardening of the earth's crust.

CHAPTER THIRTY-FIVE

The Spiral of Creation

The findings advanced in previous chapters can be brought together schematically in the diagram below:

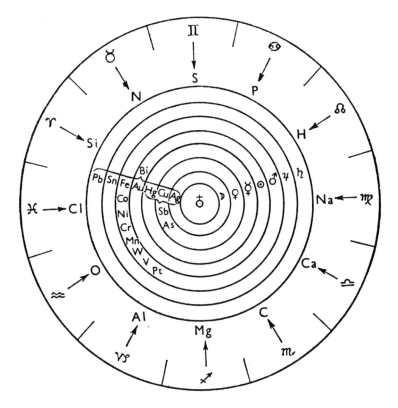

Fig. 32

Earth-substances as precipitations of cosmic forces.

The formative impulses at work in substances of the mineral earth, the hydrosphere and the atmosphere, spring from the Zodiac; whereas the impulses active in metals originate in the planets.

The question to be considered now is whether these formative impulses all work in simultaneously and equally on the earth from its cosmic surroundings, or whether there is evidence of a law which selects and guides these impulses and causes them to take effect on the earth in a certain sequence.

An ordering, governing role of this kind may readily be ascribed to the sun as the central organ of the living cosmic whole. The sun itself, or some such definite point on its path as the vernal point, which makes its round of the cosmos once in a Platonic year, could be the mediator, the regulator of the cosmic impluses that gave the earth its form. The vernal point might serve as the outlet through which a formative impulse streams in to shape an earth always in the making.

The vernal point's advance from one constellation to another has marked a corresponding advance from culture to culture in the course of history. This point now lies in the constellation Pisces. When it was progressing through Aries, the Graeco-Roman epoch ran its course. It moved through Taurus during the Egypto-Chaldean culture, when the leadership principle was worshipped under the image of the Bull. Earlier still came the Gemini culture of the Persians, with its central emphasis on the antithesis of light and darkness, Ormuzd and Ahriman. Still prior to this there was the ancient Indian civilization, of which the Vedas are only a late echo. The constellation Cancer ruled this period. As smaller rhythms are always part of similar, larger rhythms, we may assume that the position of the vernal point in periods preceding these not only left its mark on human spiritual development, but decisively affected matter in the forming of substances.

It takes a Platonic year (25,920 sun years) for the vernal point to complete one round of the cosmos. If, in its journeying, it is to transmit and mediate the formative impulses radiating in upon the earth from the Zodiac, the result of these two inter-acting movements is an inward-turning spiral, which would have to wind its way through all the planetary spheres before reaching earth.

Indeed, a spiral tendency is the evolutionary pattern more or less clearly present wherever life has left its imprint or is in process of developing. An example is found in the morphology of plants. It is not hard to discover the spiral in the arrangement of leaves around a stalk or in rose-petal patterns.

A spiral is even hidden in the acceleration pattern of a falling body. An even rate of travel on a spiral path, seen in side elevation, produces pendulum-like movements of a length and duration that decrease in proportion as their frequency rises. Pendular motion is governed by laws very similar to those obtaining in free fall.

A physician discovered that the law of falling bodies applies also to embryonic development in a striking way. The lengthwise growth of the embryo is at first almost imperceptible. If measurements are taken at regular intervals during pregnancy, however, one finds that the initial minimal growth-rate speeds up increasingly, and just before birth makes a tremendous spurt. This happens also with free-falling objects. The acceleration of a falling stone can be calculated by the velocity formula $V = \frac{g}{2} t^2$

In the 1st second a stone falls   5 metres
In the 2nd second a stone falls 20 metres
In the 3rd second a stone falls 45 metres
In the 4th second a stone falls 80 metres
In the 5th second a stone falls 125 metres
In the 6th second a stone falls 180 metres
In the 7th second a stone falls 245 metres
In the 8th second a stone falls 320 metres
In the 9th second a stone falls 405 metres

We see here that the rate of fall is almost one hundred times as great in the ninth second as in the first. The rate of increase in the size of an embryo, determined in monthly measurements, is roughly proportional to that of falling bodies. We might say that new-born human beings fall to earth out of the cosmos according to the law of falling bodies. We can at least not lightly dismiss the possibility of a connection between the spiral pattern and creative processes as witnessed in the development of man, earth and universe.

As the great spiral of creative evolution rolls in from the cosmos towards the earth, passing through the various planetary spheres as time goes on, it transmits the formative impulses radiating from the zodiacal constellations to the earth. This zodiacal radiation would be the sole origin of substances were it not for the fact that the mediating, ordering, creative force is

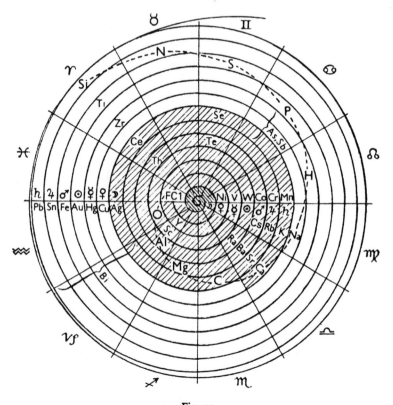

Fig. 33

The spiral of creation.

always passing through one or other of the planetary spheres. The planet thus tempers the formative impulses of the constellation.

On its next round, the in-spiralling path transmits the same constellation, but has meanwhile entered a planetary sphere closer to the earth. This gives a different tempering to the creative action of the constellation. That is why certain earth-substances resemble each other so closely. They are of the same macrocosmic origin, but they also differ as a result of differing planetary influences.

It is a universal law that smaller cycles are part of larger cycles, lesser rhythms part of greater ones. These phenomena therefore have a certain periodicity. There are good reasons to assume that one of the Platonic years began with Aries, the constellation in which the vernal point was located at the start of our era. It is natural to think of Aries as the starting point of the whole cycle of creation, to picture its evolutionary spiral coming from cosmic infinitude and entering the realm of our planetary system at the Saturn sphere and in the Aries segment of the Zodiac.

With the passage of the spiral through the Saturn sphere, the formative impulse of Aries was stirred to action. Had this been able to radiate in upon the earth without any interference, it would have produced a substance that was a simple fixation of the Aries process. But the fact that the stimulating force was also in the Saturn sphere meant that a Saturnian tempering was given to the Aries impulse. Earthly silica was the result. It is easy to see the Saturnian influence in the properties of silica. One has only to recall its surface-forming tendency and its connection with the sense organs, where Aries and Saturn forces work together.

In its further progress, the spiral then passes through the Mars sphere. As it enters the Aries region on its second round, it again releases the forces of this region. But this time they receive a Martian tempering, with the resultant earthly creation of titanium (titanium dioxide), a substance very closely resembling silica. Like silica, it is a component of primeval rock. Mineralogists call it rutile. It is a younger brother of silica; physically and chemically it behaves almost exactly like silica. It betrays the Mars impulse in its crystalline form, for it does not crystallize in the aristocratic columns and pyramids typical of rock crystal, but in needle form. The 'blood relationship' of the two shows up in the fact that rock crystal often encloses sheaves of rutile needles. And titanium has a noteworthy relationship to Mars through its occurrence in certain iron ores, especially iron rose. And – so wonderfully do these things fit together – the fragrant red garden roses, which are the offspring of Mars' creativity in the plant world, contain titanium in their ash.

Just as the Aries process is modified by the Mars sphere to form titanium, Mercury modifies it to form zirconium, followed by cerium and thorium. Thus a group of related earthly substances comes into being as the original formative impulses undergo modification by the various planetary spheres.

The silica minerals, then, are products of the Aries impulse. Sulphur and its relatives, selenium and tellurium, derive from Gemini; sodium and the other alkalis – lithium, potassium, rubidium and caesium – from Virgo; lime and the mineral alkalis – strontium, barium and radium – from Libra; aluminium and its brothers – scandium, yttrium and several other rarer minerals – from Capricorn; and the halogens, finally, from Pisces.

These groups of related substances recall the Periodic Table, where they are also arranged in groups. The Periodic Table might justifiably be thought of as the final expression of the creative cosmic symphony, which can be experienced in a more living way as it sounds through the spiral of creation. The Periodic Table could be described as a static abstraction. In the creative spiral, on the other hand, the succession of events in time is expressed.

Every year, as we know, the vernal point drops back a little, so that in the course of about two thousand years it passes through one constellation in the opposite direction to the sun's course. Applying the law of the periodicity of smaller cycles within larger ones here, we can assume that in the course of each Platonic year the formative impulse released by the creative spiral drops back one constellation. Thus, after the Saturn sphere had tempered the Aries impulse and formed silica, the next creative impulse would be released in the constellation Taurus before the end of that Platonic year. Meanwhile the spiral has entered the Jupiter sphere. The formative impulse of Taurus tempered by Jupiter leads to the formation of nitrogen. The Jupiter aspect of nitrogen will be evident in the light of our description of tin.

The creative spiral then enters the Mars sphere, and again, before the end of the next Platonic year, releases those Gemini forces that lead on earth to the forming of sulphur. Similarly, Cancer, tempered by the sun sphere, produces phosphorus; Leo tempered by Mercury, hydrogen; Virgo tempered by Venus, sodium; Libra tempered by the moon sphere, calcium.

This would constitute creation in its entirety. But we must attribute to the moon sphere the same reflective capacity that we found to be a property of the moon and saw in the physical and chemical aspects of silver. As the spiral of creation nears the earth, it passes through the moon sphere, at the same time making a transit of the reflected spheres of the other planets. Thus we find not only a moon aspect in calcium, but a Saturnian as well. It is a peculiarity of lime that silver and lead are often found embedded in it – a fact that was brought out in Chapter Nineteen. And we see the formative impulses of Saturn and Libra uniting in the calcium framework of our bones.

It may be asked here why the Saturn sphere is the one reflected, when the laws of reflection would lead us to expect it to be Venus. But on closer study it seems that the reflection here is more in the nature of a repetition, a recapitulation of an entire cosmic pattern, in the sense that wave follows wave.

After completing its transit of the reflected Saturn sphere, the creative spiral enters the reflected sphere of Jupiter, and in conjunction with Scorpio releases the forces that form carbon. It is easy to grasp the connection of Jupiter with carbon if we recall that its structure of chains and rings makes possible the fashioning of organic substance. In a picture – carbon's relatedness to tin enables it to keep on soldering itself together. The Jupiter process is a sculptor both in the fluid element and in our thought-life, where the power of association links concepts in a 'train of thought', somewhat as carbon forms chains and rings in the chemical structuring of substances.

To continue; the spiral creates magnesium under the influence of Sagittarius working in conjunction with the reflected Mars sphere, and aluminium under Capricorn working with the sun-sphere. Again, it is not difficult to discern the sun aspect in Capricorn's aluminium process, which we may remember as the harmonizing, balancing force at work in gem-stones, especially tourmaline, while recalling also that aluminium joins forces with the central directive power of the universe found in the activity of gold.

Similarly, oxygen comes into being under the influence of Aquarius in the reflected Mercury-sphere; and fluorine and the halogens, finally, originate in the reflected Venus sphere under the influence of Pisces.

Again, as fig. 33 shows, the places of origin of these substances form a spiral that runs counter to the course of the creative spiral, just as the path of the vernal point in the ecliptic moves counter to the direction of the sun.

This pattern takes account of the time element in creative evolution. It pictures earth's development from the earliest phase of material evolution, beginning with a light and air phase, passing through a fluid state, and only at the very end of creation becoming solid.

The Periodic Table brings out some of the same facts as the spiral, but by no means all. There is bound to be a difference between the living picture drawn in these pages and an abstract system of numerical tabulation.

The metals, for which there seems to be no suitable place in the Periodic Table, appear in the spiral as materialized planetary processes. It would seem natural to put the 'brothers of iron' that spring from modifications of the Mars process in the various reflected planetary spheres, since the modifying forces that produced them were terrestrial ones that affected the moon's reflecting realm.

The emergence of the material world, however, is not to be thought of as having come about through a kind of short-circuit between earth and cosmos. Cosmic order weaves and pulsates through the whole of creation, descending step by step to the mineral phase of earthly substance. The creative spiral permeates every level of events – the spiritual, the psychological, the biological, the mineral.

'Wherever lime or magnesium are found, a plant must have lived and produced these substances,' says Herzeele. 'The first milligramme of lime is no older than the first plant.' Creative impulses always act first in the organic realm, in organisms. But we must conceive of organisms as far less clearly defined in those early days – still as processes, not yet as isolated single forms.

We must also take into account the fact that the impulses of evolution were not always able to proceed in a straight line, without interruption. The spiral of creation is to be conceived rather as a framework which can help us to live into the evolution of substances. It is not to be taken as a hard and fast system in every detail, but rather as an attempt to picture the dynamics of emergent material evolution.

CHAPTER THIRTY-SIX

Conclusion

The sun makes music as of old
Among the rival spheres of heaven.
Goethe (translation by Shelley)

In these words, Goethe brings us an echo of those very ancient times when human beings still actually heard the sounding harmonies of the cosmos and felt how by them heaven and earth were formed. Plato, one of the last men able to experience this ground-tone of creation, called it the 'harmony of the spheres'.

The cosmos resounds. There is a twelvefold sounding from the zodiacal constellations, a sevenfold sounding from the planetary spheres.

The cosmic 'Word' described by St. John as having 'made everything that was made', draws the consonantal framework of its body from the Zodiac, the sounding vowels of its all-permeating music from the planetary spheres. These two elements combine in ever-new varieties of ways to fashion the world thus far created and still in process of becoming.

Earth, like heaven, is shaped by the musical ordering power of this cosmic Word, which reaches into matter itself and gives it patterns of coherence. Earth is the materialized cosmic Word, 'the end of God's path'.

When we go out into nature and look at trees and flowers, stones and mountains, veins of ore, trying to grasp all we see in a way that will lead us towards the creative archetypes, we can perhaps catch a glimpse here and there of what lies beneath the surface. This means beginning to read the divine cosmic Word again – a reading that frees it from captivity in matter.

The word chemistry comes from the Egyptian 'chemi', which originally meant 'dark earth'. Later on, it came to mean the science of obscure, secret things – 'occult' as opposed to manifest, accessible knowledge. The ancients could experience cosmic forces as directly manifest, whereas the earth, as the finished work of creation, seemed to them, by comparison, hidden and impenetrable. Chemistry, therefore, was knowledge of the 'hidden'.

Nowadays the situation is just the opposite. Knowledge of divine reality is lost; earth has become the object of our researches. Physics, chemistry and the other natural sciences have made the things of earth familiar knowledge. A science that dealt with divine reality would be called 'occult' in the sense that the facts and foundations of such a study are not 'manifest' to people of the present time.

Yet if we are to strive for a fulness of knowledge, we must seek for a scientific outlook which will embrace both halves of reality, the heavenly as well as the terrestrial.

Astronomy, of course, is not a true science of the heavens in our sense; it is terrestrial science projected into cosmic space. On the other hand, the doctrines and dogmas of the various Churches no longer satisfy modern man, for they cannot bridge the gap that separates them from terrestrial science, and modern humanity wants to *know*, not simply to believe.

We have attempted to show here how it is possible, while maintaining scientific exactness, to feel one's way towards the living laws of the world; to find and unravel the threads which lead from earthly matter to its origin in the creative spheres of the cosmos. This book was meant to be a contribution to discerning the pulse-beat of the cosmic organism. The author's hope is that it may encourage further progress, through which these unfinished first steps will be justified.

Recommended Further Reading

Kolisko, E. & L., *Agriculture of Tomorrow*, Second Edition, 1978, Kolisko Archive Publications, England

Fyfe, Agnes, *Moon and Plant Growth*, Capillary Dynamic Studies, Society for Cancer Research, Arlesheim, Switzerland

Kolisko, E., *Nutrition No. 1* (3 lectures)
  Nutrition No. 2 (2 lectures), Kolisko Archive Publications, England

Kolisko, E., *Nutrition and Agriculture*, Kolisko Archive Publications, England

Kolisko, E., *Twelve Groups of Animals*, Kolisko Archive Publications, England

Rudolf Hauschka
Nutrition
A Holistic Approach

Today's understanding of nutrition is based largely on physical, chemical considerations and analysis. Hauschka takes a radically different approach, viewing matter—and food in particular—as having a spiritual aspect. From this holistic perspective he presents a new, practical approach to nutrition.

This classic work, reprinted in its original form, is the result of Dr Hauschka's many years' research at the Ita Wegman Clinic in Arlesheim, Switzerland. Through active experimentation, Hauschka found fresh insight into the principles of digestion, which enabled him to evolve a system of nutrition suitable for the present day.

In an age of mass food production, Hauschka considers one of the most neglected aspects of nutrition—the *quality* of food. He discusses aspects of food which can be measured by conventional scientific means, as well as aspects which defy quantification by the usual methods. He relates these findings to a historical survey of food cultivation, preparation and preservation, and to the question of the chemically treated foods of today.

In the present climate of food scares and concerns—BSE, foot and mouth disease, genetic modification, chemical poisoning, etc.—Hauschka's book takes on a new relevance, adding a significant contribution to the current debate. Also included are concise dietary suggestions by Dr Margarethe Hauschka for healthy as well as sick people.

This book is a companion volume to *The Nature of Substance*.

248pp; 1 85584 117 7; £12.95

Theodor Schwenk

Sensitive Chaos
The Creation of Flowing Forms in Water and Air

'From space the Earth is seen as a water planet, less than 30 per cent is land. Our sister planets Mars and Venus were made of the same stuff when they started but are now drier than any conceivable desert on Earth. We know that without water there can be no life, but also it is true that without life there can be no water. In *Sensitive Chaos* Theodor Schwenk teaches us about this wonderful connection between water and life. So movingly and well told is his tale that you will not want to put the book down until the end.'
James Lovelock, scientist and author of *Ages of Gaia* and *Gaia—A New Look at Life on Earth*

Why does water, in streams and rivers, always take a winding course? Do common principles and rhythms underlie the movement of water, whether it be in the sea or in a plant—or in the blood of a human being?

The laws revealed in the subtle patterns of water in movement are shown in this thought-provoking work to be the same as those perceptible in the shaping of bones, muscles and myriads of other forms in nature. Lavishly illustrated, *Sensitive Chaos* shows the unifying forces which underlie all living things, and observes such phenomena as the flight of birds; the movement of fish; the formation of internal organs such as the heart, eye and ear; air patterns in musical instruments; the formation of mountain ranges and river deltas; sand patterns on the beach; weather and space patterns; and even the formation of the human embryo.

288pp (72 b&w plates); 1 85584 055 3; £18.95